Machine Learning in Biotechnology and Life Sciences

Build machine learning models using Python and deploy them on the cloud

Saleh Alkhalifa

BIRMINGHAM—MUMBAI

Machine Learning in Biotechnology and Life Sciences

Publishing Product Manager: Ali Abidi

Senior Editor: David Sugarman

Content Development Editor: Nathanya Dias

Technical Editor: Rahul Limbachiya

Copy Editor: Safis Editing

Project Coordinator: Aparna Ravikumar Nair

Proofreader: Safis Editing

Indexer: Pratik Shirodkar

Production Designer: Joshua Misquitta

Marketing Coordinator: Abeer Riyaz Dawe

First published: January 2022

Production reference: 1221221

Published by Packt Publishing Ltd.

Livery Place

35 Livery Street

Birmingham

B3 2PB, UK.

ISBN 978-1-80181-191-0

www.packt.com

Contributors

About the author

Saleh Alkhalifa is a data scientist and manager in the biotechnology industry with 4 years of industry experience working and living in the Boston area. With a strong academic background in the applications of machine learning for discovery, prediction, forecasting, and analysis, he has spent the last 3 years developing models that touch all facets of business and scientific functions.

I would like to dedicate this book to my parents, Esam Alkhalifa and Anna Letz, without whose motivation and support I would not be the scientist I am today.

About the reviewers

Indraneel Chakraborty is a data science enthusiast with an open mindset and a passion for solving data problems with code and building cloud-based data apps. He has both academic and industrial skill sets that include experience in the curation and analysis of clinical trials registry data for insights into policy research and also experience working with biomedical data providing ML/AI-ready solutions. He enjoys coding in Python and R, with lots of Googling, of course!

Dr Neha Kalla has a PhD in biotechnology from Banasthali University, India. During her PhD, she spent time working as a project assistant in an Indo-Danish project and also worked at the Centre for Conservation and Utilization of BGA, Indian Agricultural Research Institute, India. After completing her PhD, she developed associations with some of the most renowned professors and scientists across India and at Cambridge University. Following a long stint in research and academia, she decided to proceed with her career as a data scientist. She is presently working as a senior data scientist in the sphere of AI, offering machine learning-based retail solutions in Germany.

Table of Contents

3
Getting Started with SQL and Relational Databases

4
Visualizing Data with Python

Section 2: Developing and Training Models

8
Understanding Deep Learning

9
Natural Language Processing

Section 3: Deploying Models to Users

Index

Other Books You May Enjoy

Preface

We have seen major changes in the field of machine learning in the last few years that have impacted our daily lives and the way business decisions are made. If there is one thing that the biotechnology and life sciences industries have in abundance, it is their never-ending sources of data. As we move toward more data-driven models, the intersection of life sciences and machine learning has seen unprecedented growth, uncovering vast quantities of information and hidden patterns giving companies major competitive advantages.

Over the course of this book, we will touch on some of the most important elements of machine learning from both a supervised and unsupervised perspective. We will not only learn to develop and train robust models, but also deploy them in the cloud using AWS and GCP, allowing us to make them immediately available for end users.

Who this book is for

This book specifically caters to scientific professionals in both academia and industry looking to transcend to the data science domain. Individual contributors and managers alike who are already established within the pharmaceutical, life sciences, and biotechnology sectors will find this book not only useful, but immensely applicable to current-day projects. Although an introduction to Python and machine learning is provided, a basic understanding of Python programming and a beginner-level background in data science conjunction is recommended to get the most out of this book.

What this book covers

Chapter 1, *Introducing Machine Learning for Biotechnology*, provides a brief introduction to the field of biotechnology and some of the areas in which machine learning can be applied, in addition to some of the technology this book will use.

Chapter 2, *Introducing Python and the Command Line*, comprises a summary of some of the must-know techniques and commands in Bash and the Python programming language, in addition to some of the most common Python libraries.

Chapter 3, *Getting Started with SQL and Relational Databases*, is where you will gain knowledge of the SQL querying language and learn how to create a remote database using MySQL and AWS RDS.

Chapter 4, *Visualizing Data with Python*, introduces you to some of the most common methods for visualizing and representing data using the Python programming language.

Chapter 5, *Understanding Machine Learning*, comprises some of the most important elements of standard machine learning pipelines, introducing you to supervised and unsupervised methods, as well as saving models for future use.

Chapter 6, *Unsupervised Machine Learning*, is where you will learn about unsupervised models and dive into clustering and dimensionality reduction methods with tutorials relating to breast cancer.

Chapter 7, *Supervised Machine Learning*, is where you will learn about supervised learning models and dive into classification and regression methods.

Chapter 8, *Understanding Deep Learning*, provides an overview of the deep learning space, where we will explore the elements of a deep learning model, as well as two tutorials relating to protein classification using Keras and anomaly detection using AWS.

Chapter 9, *Natural Language Processing*, teaches you some of the most common NLP options as we explore popular libraries and tools, in addition to two tutorials relating to clustering as well as semantic searching using transformers.

Chapter 10, *Exploring Time Series Analysis*, explores data using a time-based approach in which we break down the components of a time series dataset and develop two forecasting models using Prophet and LSTMs.

Chapter 11, *Deploying Models with Flask Applications*, provides an introduction to one of the most popular frameworks for deploying models and applications to end users.

Chapter 12, *Deploying Applications to the Cloud*, provides an introduction to two of the most popular cloud computing platforms, in addition to three tutorials allowing users to deploy their work to AWS LightSail, GCP AppEngine, and GitHub.

To get the most out of this book

To maximize the value of your time, a very basic knowledge of the Python programming language and the Bash command line is recommended. In addition, some background in the biotechnology and life sciences spheres is recommended to best understand the tutorials and use cases.

Software/hardware covered in the book	Operating system requirements
Python 3	Windows, macOS, or Linux
Jupyter Notebooks	Windows, macOS, or Linux
MySQL	Windows, macOS, or Linux
Anaconda Individual Edition	Windows, macOS, or Linux
AWS Account	Any modern-day web browser
GCP Account	Any modern-day web browser
Git	Windows, macOS, or Linux

If you are using the digital version of this book, we advise you to type the code yourself or access the code from the book's GitHub repository (a link is available in the next section). Doing so will help you avoid any potential errors related to the copying and pasting of code.

Download the example code files

You can download the example code files for this book from GitHub at `https://github.com/PacktPublishing/Machine-Learning-in-Biotechnology-and-Life-Sciences`. If there's an update to the code, it will be updated in the GitHub repository.

We also have other code bundles from our rich catalog of books and videos available at `https://github.com/PacktPublishing/`. Check them out!

Download the color images

We also provide a PDF file that has color images of the screenshots and diagrams used in this book. You can download it here: `https://static.packt-cdn.com/downloads/9781801811910_ColorImages.pdf`.

Conventions used

There are a number of text conventions used throughout this book.

`Code in text`: Indicates code words in the text, database table names, folder names, filenames, file extensions, pathnames, dummy URLs, user input, and Twitter handles. Here is an example: "Mount the downloaded `WebStorm-10*.dmg` disk image file as another disk in your system."

A block of code is set as follows:

```
from sklearn.preprocessing import StandardScaler
scaler = StandardScaler()
X_scaled = scaler.fit_transform(dfx.drop(columns =
["annotation"]))
```

When we wish to draw your attention to a particular part of a code block, the relevant lines or items are set in bold:

```
>>> heterogenousList[0]
dichloromethane
>>> heterogenousList[1]
3.14
```

Any command-line input or output is written as follows:

```
$ mkdir machine-learning-biotech
```

Bold: Indicates a new term, an important word, or words that you see on screen. For instance, words in menus or dialog boxes appear in **bold**. Here is an example: "Select **System info** from the **Administration** panel."

> **Tips or Important notes**
> Appear like this.

Get in touch

Feedback from our readers is always welcome.

General feedback: If you have questions about any aspect of this book, email us at customercare@packtpub.com and mention the book title in the subject of your message.

Errata: Although we have taken every care to ensure the accuracy of our content, mistakes do happen. If you have found a mistake in this book, we would be grateful if you would report this to us. Please visit www.packtpub.com/support/errata and fill in the form.

Piracy: If you come across any illegal copies of our works in any form on the internet, we would be grateful if you would provide us with the location address or website name. Please contact us at copyright@packt.com with a link to the material.

If you are interested in becoming an author: If there is a topic that you have expertise in and you are interested in either writing or contributing to a book, please visit authors.packtpub.com.

Share your thoughts

Once you've read *Machine Learning in Biotechnology and Life Sciences*, we'd love to hear your thoughts! Scan the QR code below to go straight to the Amazon review page for this book and share your feedback.

https://packt.link/r/1-801-81191-1

Your review is important to us and the tech community and will help us make sure we're delivering excellent quality content.

Section 1: Getting Started with Data

This section describes the basics of Python, SQL, and translating raw data into meaningful visualizations and representations as the first step of a strong data science project. Novice students generally find themselves overwhelmed by the vast amount of data science content found on the internet or in print. This book remedies this issue by focusing on the most important and valuable must-know elements for getting started in the field.

This section comprises the following chapters:

- *Chapter 1, Introducing Machine Learning for Biotechnology*
- *Chapter 2, Introducing Python and the Command Line*
- *Chapter 3, Getting Started with SQL and Relational Databases*
- *Chapter 4, Visualizing Data with Python*

1

Introducing Machine Learning for Biotechnology

How do I get started? This is a question that I have received far too frequently over my last few years as a data scientist and consultant operating in the technology/biotechnology sectors, and the answer to this question never really seemed to change from person to person. My recommendation was generally along the lines of learning **Python** and **data science** through online courses and following a few tutorials to get a sense of how things worked. What I found was that the vast majority of scientists and engineers that I have encountered, who are interested in learning data science, tend to get overwhelmed by the large volume of resources and documentation available on the internet. From *Getting Started in Python* courses to *Comprehensive Machine Learning* guides, the vast majority of those who ask the question *How do I get started?* often find themselves confused and demotivated just a few days into their journey. This is especially true for scientists or researchers in the lab who do not usually interact with code, algorithms, or predictive models. Using the Terminal command line for the first time can be unusual, uncomfortable, and – to a certain extent – terrifying to a new user.

This book exists to address this problem. This is a one-stop shop to give **scientists**, **engineers**, and everyone in-between a fast and efficient guide to getting started in the beautiful field of data science. If you are not a coder and do not intend to be, you have the option to read this book from cover to cover without ever using Python or any of the hands-on resources. You will still manage to walk away with a strong foundation and understanding of machine learning and its useful capabilities, and what it can bring to the table within your team. If you are a coder, you have the option to follow along on your personal computer and complete all the tutorials we will cover. All of the code within this book is inclusive, connected, and designed to be fully replicable on your device. In addition, all of the code in this book and its associated tutorials is available online for your convenience. The tutorials we will complete can be thought of as blueprints to a certain extent, in the sense that they can be recycled and applied to your data. So, depending on what your expectations of the phrase *getting started* are, you will be able to use this book effectively and efficiently, regardless of your intent to code. So, how do we plan on getting started?

Throughout this book, we will introduce concepts and tutorials that cater to problems and use cases that are commonly experienced in the technology and biotechnology sectors. Unlike many of the courses and tutorials available online, this book is well-connected, condensed, and chronological, thus offering you a fast and efficient way to get up to speed on data science. In under 400 pages, we will introduce the main concepts and ideas relating to Python, SQL, machine learning, deep learning, natural language processing, and time-series analysis. We will cover some popular approaches, best practices, and important information every data scientist should know. In addition to all of this, we will not only put on our data scientist hats to train and develop several powerful predictive models, but we will also put on our data engineer hats and deploy our models to the cloud using **Amazon Web Services** (**AWS**) and **Google Cloud Platform** (**GCP**). Whether you are planning to bring data science to your current team, train and deploy the models yourself, or start interviewing for data scientist positions, this book will equip you with the right tools and resources to start your new journey, starting with this first chapter. In the following sections, we will cover a few interesting topics to get us started:

- Understanding the biotechnology field
- Combining biotechnology and machine learning
- Exploring machine learning software

With that in mind, let's look at some of the fun areas within the field of biotechnology that are ripe for exploration when it comes to machine learning.

Understanding the biotechnology field

Biotechnology, as the name suggests, can be thought of as the area of technological research relating to biology when it comes to living organisms or biological systems. First coined in 1919 by Karoly Ereky, the *father* of biotechnology, the field traditionally encompassed the applications of living organisms for commercial purposes.

Some of the earliest applications of biotechnology throughout human history include the process of fermenting beer, which dates as far back as 6,000 BC, or preparing bread using yeast in 4,000 BC, or even the development of the earliest viral vaccines in the 1700s.

In each of these examples, scientific or engineering processes utilized biological entities to produce goods. This concept was true then and had remained just as true throughout human history. Throughout the 20th century, major innovative advancements were made that changed the course of mankind for the better. In 1928, Alexander Fleming identified a mold that halted the replication of bacteria, thus leading to penicillin – the first antibiotic. Years later, in 1955, Jonas Salk developed the first polio vaccine using mammalian cells. Finally, in 1975 one of the earliest methods for the development of monoclonal antibodies was developed by George Kohler and Cesar Milstein, thus reshaping the field of medicine forever:

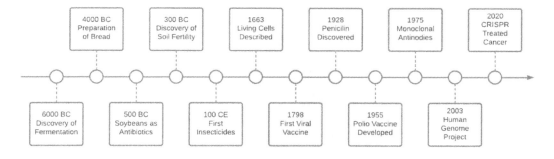

Figure 1.1 – A timeline of a few notable events in the history of biotechnology

Toward the end of the 20th century and the beginning of the 21st century, the field of biotechnology expanded to cover a diverse bevy of sub-fields, including genomics, immunology, pharmaceutical treatments, medical devices, diagnostic instruments, and much more, thus steering its focus away from its agricultural applications and more on human health.

Success in Biotech Health

Over the last 20 years, many life-changing treatments and products have been approved by the FDA. Some of the industry's biggest blockbusters include Enbrel® and Humira®, monoclonal antibodies for treating rheumatoid arthritis; Keytruda®, a humanized antibody for treating melanoma and lung cancer; and, finally, Rituxan®, a monoclonal antibody for treating autoimmune diseases and certain types of cancer. These blockbusters are but a sample of the many significant advances that have happened in the field over the past few decades. These developments contributed to creating an industry that's larger than many countries on Earth while changing the lives of millions of patients for the better.

The following is a representation of a monoclonal antibody:

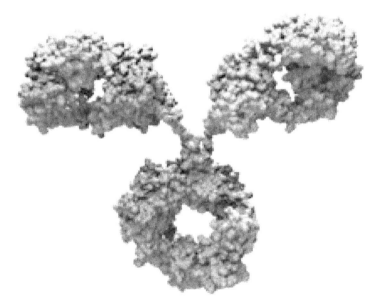

Figure 1.2 – A 3D depiction of a monoclonal antibody

The biotechnology industry today is flourishing with many new and significant advances for treating illnesses, combatting diseases, and ensuring human health. However, with the space advancing as quickly as it is, the discovery of new and novel items is becoming more difficult. A great scientist once told me that advances in the biopharmaceutical industry were once made possible by pipettes, and then they were made possible by automated instruments. However, in the future, they will be made possible by **Artificial Intelligence (AI)**. This brings us to our next topic: **machine learning**.

Combining biotechnology and machine learning

In recent years, scientific advancements in the field, boosted by applications of machine learning and various predictive technologies, have led to many major accomplishments, such as the discovery of new and novel treatments, faster and more accurate diagnostic tests, greener manufacturing methods, and much more. There are countless areas where machine learning can be applied within the biotechnology sector; however, they can be narrowed down to three general categories:

- **Science and Innovation**: All things related to the research and development of products.

- **Business and Operations**: All things related to processes that bring products to market.

- **Patients and Human Health**: All things related to patient health and consumers.

These three categories are essentially a product pipeline that begins with scientific innovation, where products are brainstormed, followed by business and operations, where the product is manufactured, packaged, and marketed, and finally the patients and consumers that utilize the products. Throughout this book, we will touch on numerous applications of machine learning as they relate to these three fields within the various tutorials that will be presented. Let's take a look at a few examples of applications of machine learning as they relate to these areas:

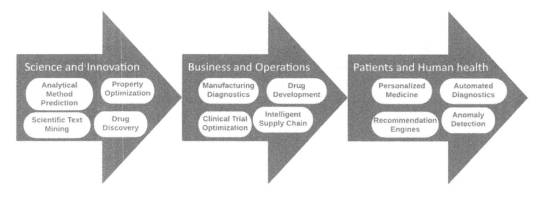

Figure 1.3 – The development of a product highlighting areas where AI can be applied

Throughout the life cycle of a given product or therapy, there are numerous areas where machine learning can be applied – the only limitation is the existence of data to support the development of a new model. Within the scope of **science and innovation**, there have been significant advances when it comes to predicting molecular properties, generating molecular structures to suit specific therapeutic targets, and even sequencing genes for advanced diagnostics. In each of these examples, AI has been – and continues to be – useful in aiding and accelerating the research and development of new and novel products. Within the scope of **business and operations**, there are many examples of AI being used to improve processes such as intelligently manufacturing materials to reduce waste, natural language processing to extract insights from scientific literature, or even demand forecasting to improve supply chain processes. In each of these examples, AI has been crucial in reducing costs and increasing efficiency. Finally, when it comes to **patients and health**, AI has proven to be pivotal when it comes to recruiting people for and shaping clinical trials, developing recommendation engines designed to avoid drug interactions, or even faster diagnoses, given a patient's symptoms. In each of these applications, data was obtained, used to generate a model, and then validated.

The applications of AI we have observed thus far are only a few examples of the areas where powerful predictive models can be applied. In almost every process throughout the cycle where data is available, a model can be prepared in some way, shape, or form. As we begin to explore the development of many of these models in various areas throughout this process, we will need a few software-based tools to help us.

Exploring machine learning software

Before we start developing models, we will need to few tools to help us. The good news is that regardless of whether you are using a **Mac**, **PC**, or **Linux**, almost everything we will use is compatible with all platforms. There are three main items we will need to install: a language to develop our models in, a database to store our data in, and a cloud computing space to deploy our models in. Luckily for us, there is a fantastic technology stack ready to support our needs. We will be using the Python programming language to develop our models, MySQL to store our data, and AWS to run our cloud computing processes. Let's take a closer look at these three items.

Python (programming language)

Python is one of the most commonly used programming languages and sought-after skills in the data science industry today. It was first developed in 1991 and is regarded today as the most common language for data science. For this book, we will be using Python 3.7. There are several ways you can install Python on your computer. You can install the language in its standalone form from `Python.org`. This will provide you with a Python interpreter in its most basic form where you can run commands and execute scripts.

An alternative installation process that would install Python, pip (a package to help you install and manage Python libraries), and a collection of other useful **libraries** can be done by using **Anaconda**, which can be retrieved from `anaconda.com`. To have a working version of Python and its associated libraries on your computer as quickly as possible, using Anaconda is highly recommended. In addition to Python, we will need to install libraries to assist in a few areas. Think of libraries as nicely packaged portions of code that we can import and use as we see fit. Anaconda will, by default, install a few important libraries for us, but there will be others that we will need. We can install those on-the-go using pip. We will look at this in more detail in the next chapter. For the time being, go ahead and install Anaconda on your computer by navigating to the aforementioned website, downloading the installation that best matches your machine, and following the installation instructions provided.

MySQL (database)

When handling vast quantities of information, we will need a place to store and save all of our data throughout the analysis and preprocessing phases of our projects. For this, we will use MySQL, one of the most common relational databases used to store and retrieve data. We will take a closer look at the use of MySQL by using SQL. In addition to the MySQL relational database, we will also explore the use of DynamoDB, a non-relational and NoSQL database that has gained quite a bit of popularity in recent years. Don't worry about getting these setups right now – we will talk about getting them set up later on.

AWS and GCP (Cloud Computing)

Finally, after developing our machine learning models in Python and training them using the data in our databases, we will deploy our models to the cloud using both **Amazon Web Services** (**AWS**), and **Google Cloud Platform** (**GCP**). In addition to deploying our models, we will also explore a number of useful tools and resources such as Sagemaker, EC2, and AutoPilot (AWS), and Notebooks, App Engine, and AutoML (GCP).

Summary

In this chapter, we gained a quick understanding of the field of biotechnology. First, we looked at some historical facts as they relate to the field, as well as some of the ways this field has been reshaped into what it looks like today. Then, we explored the areas within the field of biotechnology that are most impacted by machine learning and AI. Finally, we explored some of the most common and basic machine learning software you will need to get started in the field.

Throughout this book, Python and SQL will be the main languages we will use to develop all of our models. We will not only go through the specific instructions of how to install each of these requirements, but we will also gain hands-on knowledge throughout the many examples and tutorials within this book. AWS and GCP will be our two main cloud-based platforms for deploying all of our models, given their commonality and popularity among data scientists.

In the next chapter, we'll introduce the Python command line. With that in mind, let's go ahead and get started!

2

Introducing Python and the Command Line

When walking into a coffee shop, you will almost immediately notice three types of people: those socializing with others, those working on projects, and those who code. Coders can easily be spotted by the black background and white letters on their computer screens – this is known as the **command line**. To many, the command line can look fierce and intimidating, but to others, it is a way of life.

One of the most essential parts of conducting any type of data science project is the ability to effectively navigate directories and execute commands via the Terminal command line. The command line allows users to find files, install libraries, locate packages, access data, and execute commands in an efficient and concise way. This chapter is by no means a comprehensive overview of all the capabilities the command line has, but it does cover a general list of essential commands every data scientist should know.

In this chapter, we'll cover the following specific topics:

- Introducing the command line
- Discovering the **Python** language

- Tutorial – getting started in Python
- Tutorial – working with Rdkit and BioPython

Technical requirements

In this chapter, we will use the **Terminal** command line, which can be found in the `Applications` folder (**macOS**), or **Command Prompt**, which can be found in the Start menu (**Windows PC**). Although the two are equivalent in functionality, the syntax behind some of the commands will differ. If you are using a PC, you are encouraged to download **Git for Windows** (`https://git-scm.com/download/win`), which will allow you to follow along using the **Bash** command line. As we begin to edit files in the command line, we will need an editor called **Vim**. Most Mac users will have Vim preinstalled on their systems. PC users are encouraged to download Vim from their website (`https://www.vim.org/download.php`).

In addition, we will be exploring Python using the **Anaconda** distribution. We will go over getting this downloaded on your system soon. The process of installing Anaconda is nearly identical for both Mac and PC users, and the execution of Python code is nearly identical as well.

Throughout this book, the code you see will also be available for you on **GitHub**. We can think of GitHub as a space where code can live, allowing us to maintain versioning, make edits, and share our work with others. As you follow along in this chapter, you are encouraged to refer to the associated GitHub repository, which can be found at `https://github.com/PacktPublishing/Machine-Learning-in-Biotechnology-and-Life-Sciences`.

Introducing the command line

The command line is available for **Mac**, **PC**, and **Linux**. While the following examples were executed on a Mac, very similar functionality is also applicable on a PC, but with a slightly different syntax.

You can begin the process by opening the command line (known as **Terminal** on a Mac and **Command Prompt** on a PC). Opening Terminal will usually, by default, bring you to what's known as your **home directory**. The text you first see will specify your username and the name of your system, separated by the @ symbol. Let's look at some basic commands.

In order to identify the path of your current (working) directory, you can use the pwd command:

```
$ pwd
```

This will return the exact directory that you are currently in. In the case of my system, the returned path was as follows:

```
Users/alkhalifas
```

In order to identify the contents within this particular directory, you can use the `ls` command, which will return a list of directories and files:

```
$ ls
```

You can make a new directory within the command line by using the mkdir command, followed by the name of the directory you wish to create. For example, you can create the machine-learning-practice directory using the following command:

```
$ mkdir machine-learning-biotech
```

If you once again list the contents within this directory using the `ls` command, the new directory will appear in that list. You can navigate to that directory using the cd (that is, *change directory*) command:

```
$ cd machine-learning-practice
```

You can once again use the pwd command to check your new path:

```
$ pwd
Users/alkhalifas/machine-learning-biotech
```

In order to return to the previous directory, you can use the cd command, followed by a blank space and two periods (..):

```
$ cd ..
```

> **Important note**
> It is worth mentioning that depending on the command line you are using, the directory names can be case-sensitive. For example, entering Downloads instead of downloads can be interpreted as a different location, so this may return an error. Maintaining consistency in how you name files and directories will be key to your success in using the command line.

Creating and running Python scripts

Now that you've learned a few of the basics, let's go ahead and create our first Python application using the command line. We can create and edit a new file using **Vim**, which is a text editor. If you do not have Vim installed on your current system, you are encouraged to install it using the link provided in the *Technical requirements* section of this chapter. Once installed on your local machine, you can call the `vim` command, followed by the name of the file you wish to create and edit:

```
$ vim myscript.py
```

This will open up an empty file within the Vim editor of your command line where you can write or paste your code. By default, you will begin in the *view* mode. You can change to the *edit* mode by pressing the *I* key on your keyboard. You will notice that the status at the bottom of the Vim window has changed to - - **INSERT** - -, which means you can now add code to this file:

Figure 2.1 – The Vim window

Type (or copy and paste) the following few lines of Python code into the file, and then click the *Esc* key. You will notice that the status is no longer - - **INSERT** - - and you can no longer edit the file. Next, type :wq and press *Enter*. The *w* key will write the file and the *q* key will quit the editor. For more details on other Vim commands, see the Vim website at https://www.vim.org/:

```
# myscript.py
import datetime
now = datetime.datetime.now()
print("Hello Biotech World")
print ("The current date and time is:")
print (now.strftime("%Y-%m-%d %H:%M:%S"))
```

With that, you have now written your first Python script. We can go ahead and execute this script using the Python interpreter you installed earlier. Before we do so, let's talk about what the script will do. Starting on the first line, we will import a library known as datetime, which will allow us to determine the current date and time of our system. Next, we assign the datetime object to a variable we will call now. We will discuss objects and variables in the following section, but for the time being, think of them as variables that can be filled with values such as dates or numbers. Finally, we will print a phrase that says Hello Biotech World!, followed by a statement of the current time.

Let's give this a try:

```
$ python3 myscript.py
```

Upon executing this file, the following results will appear on your screen:

```
Hello Biotech World!
The current date and time is:
2021-05-23 18:40:21
```

In this example, we used a library called datetime, which was installed by default when you installed the Anaconda distribution. There were many others that were also installed, and many more that were not. As we progress through these projects, we will be using many of these other libraries that we can install using pip.

The previous example worked without any errors. However, this is seldom the case when it comes to programming. There will be instances in which a missing period or unclosed bracket will lead to an error. In other situations, programs will run indefinitely – perhaps in the background without your knowledge. Closing the Terminal command line generally halts running applications. However, there will be times in which closing a command-line window is not an option. To identify processes running in the background, you can use the ps (that is, process) command:

```
$ ps -ef
```

This will display a list of all running processes. The first column, UID, is the user ID, followed by the PID (process ID) column. A few columns further to the right you can see the specific names of the files (if any) that are currently active and running. You can narrow down the list using the grep command to find all those relating to Python:

```
$ ps -ef | grep python
```

If a Python script (for example, `someScript.py`) is running continuously in the background, you can easily determine the process ID by using the `grep` command, and this means you can subsequently kill the process with the `pkill` command:

```
$ pkill -9 -f someScript.py
```

This will terminate the script and free up your computer's memory for other tasks.

Installing packages with pip

One of the best resources available to manage Python libraries is the **Python Package Index (PyPI)**, which allows you to install, uninstall, and update libraries as needed. One of the main libraries we will be using is **scikit-learn** (`sklearn`), which we can install using the `pip install` command directly in the Terminal command line:

```
$ pip install sklearn
```

The package manager will then print some feedback messages alerting you to the status of the installation. In some cases, the installation will be successful, and in others, it may not be. The feedback you receive here will be helpful in determining what next steps, if any, need to be taken.

There will be instances in which one library will require another to function – this is known as a **dependency**. In some cases, `pip` will automatically handle dependencies for you, but this will not always be the case.

To identify the dependencies of a library, you can use the `pip show` command:

```
$ pip show sklearn
```

The command line will then print the name, version, URL, and many other properties associated with the given library. In some instances, the version shown will be outdated or simply not the version you need. You can either use the `pip install` command again to update the library to a more recent version, or you can select a specific version by specifying it after the library name:

```
$ pip install sklearn==0.15.2
```

As the number of packages you install begins to grow, remembering the names and associated versions will become increasingly difficult. In order to generate a list of packages in a given environment, you can use the `pip freeze` command:

```
$ pip freeze > requirements.txt
```

This command will *freeze* a list of libraries and their associated versions and then write them to a file called `requirements.txt`. This practice is common within teams when migrating code from one computer to another.

When things don't work...

Often, code will fail, commands will malfunction, and issues will arise for which a solution will not be immediately found. Do not let these instances discourage you. As you begin to explore the command line, Python, and most other code-based endeavors, you will likely run into errors and problems that you will not be able to solve. However, it is likely that others have already solved your problem. One of the best resources available for searching and diagnosing code-related issues is **Stack Overflow** – a major collaboration and knowledge-sharing platform for individuals and companies to ask questions and find solutions for problems relating to all types of code. It is highly encouraged that you take advantage of this wonderful resource.

Now that we've had a good look at the command line and its endless capabilities, let's begin to explore Python in more detail.

Discovering the Python language

There are many different computer languages that exist in the world today. **Python, R, SQL, Java, JavaScript, C++, C**, and **C#** are just a few examples. Although each of these languages is different when it comes to their syntax and application, they can be divided into two general categories: *low-level* and *high-level* languages. **Low-level** languages – such as C and C++ – are computer languages operating at the machine level. They concern themselves with very specific tasks, such as the movement of bits from one location to another. On the other hand, **high-level** languages – such as R and Python – concern themselves with more abstract processes such as squaring numbers in a list. They completely disregard what is happening at the machine level.

Before we discuss Python in more detail, let's talk about the idea of compiling programs. Most programs – such as programs written in C++ and Java – require what is known as a **compiler**. Think of a compiler as a piece of software that converts human-readable code into machine-readable code immediately before the program is started or executed. While compilers are required for most languages, they are not required for languages such as Python. Python requires what is known as an **interpreter**, which is inherently similar to a compiler in structure, but it executes commands immediately instead of translating them to machine-readable code. With that in mind, we will define Python as a *high-level*, *general-purpose*, and *interpreted* programming language. Python is commonly used for statistical work, machine learning applications, and even game and website development. Since Python is an interpreted language, it can either be used within an IDE or directly within the Terminal command line:

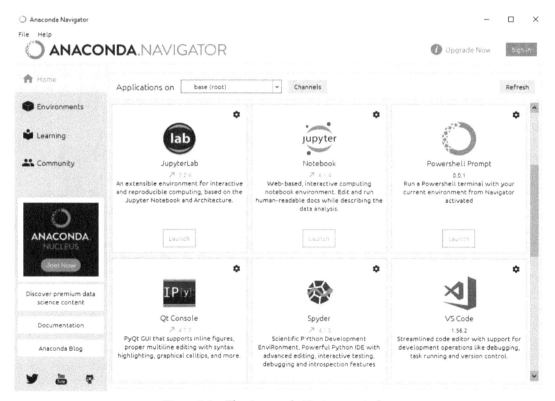

Figure 2.2 – The Anaconda Navigator window

Alternatively, you can also start the Jupyter Notebook application using the Terminal command line by typing in the following command:

```
$ jupyter notebook
```

Upon hitting enter, the same Jupyter Notebook application we saw previously should appear on your screen. This is simply a faster method of opening Jupyter Notebook.

Data types

There are many different types of data that Python is capable of handling. These can generally be divided into two main categories: **primitives** and **collections**, as illustrated in the following diagram:

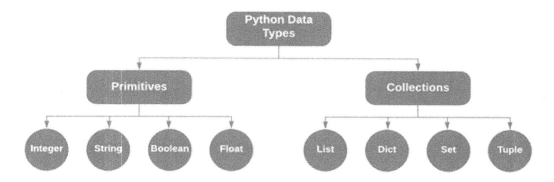

Figure 2.3 – A diagram showing Python data types

The first data type category is primitive values. As the name implies, these data types are the most basic building blocks in Python. A few examples of these include the following:

DATA TYPE	CATEGORY	EXAMPLE	ABBREVIATION
INTEGERS	Numerical	1,2,3,4,5	int
FLOATING POINTS	Numerical	1.1, 1.0, 1.2864734,	float
BOOLEANS	Truth Value	True, False	bool
STRINGS	Text	"I am a string"	str

Figure 2.4 – A table of primitive data types

The second data type category is collections. Collections are made up of one or more primitive values put together. Each type of collection has specific properties associated with it, yielding distinct advantages and disadvantages in certain conditions. A few examples of these include the following:

DATA TYPE	SPECIAL PROPERTY	EXAMPLE	ABBREVIATION
LISTS	- Order matters	[1,2,3]	list
	- Can contain anything		
	- Can be changed (mutable)		
SETS	- Contains only distinct elements	(1,2,3)	set
	- Order does not matter		
	- Can be changed (mutable)		
TUPLES	- Similar to lists	{1,2,3}	tuple
	- Cannot be changed (immutable)		
DICTIONARIES	- Most flexible	{"A": "2",	dict
	- Contain keys and values	"B": "3"}	
	- Order does not matter		
	- Can be changed (mutable)		

Figure 2.5 – A table showing different kinds of data type collections

As scientists, we are naturally inclined to organize information as best we can. We previously classified data types based on their primitive and collective nature. However, we can also classify data types based on a concept known as **mutability**. Mutability can also be thought of as *delete-ability*. Variables, such as those representing a list, hold an instance of that type. When the object is created or instantiated, it is assigned a unique ID. Normally, the type of this object cannot be changed after it is defined at runtime, however, it can be changed if it is considered *mutable*. Objects such as integers, floats, and Booleans are considered *immutable* and therefore cannot be changed after being created. On the other hand, objects such as lists, dictionaries, and sets are mutable objects and can be changed. Therefore, they are considered mutable, as you can see from *Figure 2.6*:

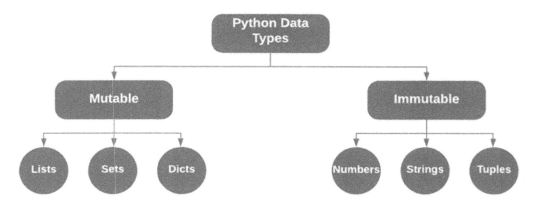

Figure 2.6 – Python data types according to their mutability

Now that we have taken a look at some of the basics, let's explore some of the more exciting areas of the Python language.

Tutorial – getting started in Python

Python is an extensive language, and any attempt to summarize its capabilities in under 10 pages would be limited. While this book is not intended to be used as a comprehensive guide to Python, we will talk through a number of the *must-know* commands and capabilities that every data scientist should be aware of. We will see the vast majority of these commands come up in the tutorials to come.

Creating variables

One of the core concepts in Python is the idea of variables. **Variables** are items that Python manipulates, with each variable having a *type*. Operators such as addition (+) or subtraction (-) can be combined with variables to create **expressions**. An example of an expression can be created using three variables. First, a value of 5 will be assigned to the x variable, and then, a value of 10 will be assigned to the y variable. The two variables (x and y), which are now representing numerical values, are considered **objects** in Python. A new variable, z, can be created to represent the sum of x and y:

```
$ python
>>> x = 5
>>> y = 10
>>> z = x + y
>>> print(z)
15
```

Variables can take on many data types. In addition to the integer values shown in the preceding code, variables can be assigned strings, floats, or even Booleans:

```
>>> x = "biotechnology"
>>> x = 3.14159
>>> x = True
```

The specific data type of a variable can be determined using the `type()` function:

```
>>> x = 55
>>> type(x)
    int
```

Data types do not need to be explicitly declared in Python (unlike other languages such as C++ or Java). In fact, Python also allows variables to be **cast** into other types. For example, we can cast an integer into a string:

```
>>> x = 55
>>> x = str(x)
>>> type(x)

    str
```

We can see from what was returned that the data is now of the string type!

Importing installed libraries

After a library has been installed, you can import the library into your Python script or Jupyter Notebook using the `import` function. You can import the library as a whole in the following way:

```
>>> import statistics
```

Alternatively, we can explicitly import all classes from the library:

```
>>> from statistics import *
```

The best way to import any library is to only import the classes that you plan to use. We can think of **classes** as standalone parts of a given library. If you plan to calculate the mean of a series of numbers, you can explicitly import the mean class from the `statistics` library:

```
>>> from statistics import mean
```

Installing and importing libraries will become second nature to you as you venture further into the data science field. The following table shows some of the most common and most useful libraries any new data scientist should know. Although not all of these will be covered within the scope of this book, they are useful to know.

LIBRARY	ABBREVIATION	EXPLANATION
PANDAS	pandas	A data-handling library known for DataFrames
NUMPY	numpy	A data-handling library known for Numpy arrays
SCIKIT-LEARN	sklearn	A machine learning algorithm repository
MATPLOTLIB	matplotlib	Used for graphing data and creating plots
NLTK	nltk	A natural language processing library
REQUESTS	requests	Importing data from websites using URLs
SEABORN	seaborn	Used for graphing data and creating plots
DATETIME	datetime	A library catered to organizing dates and times
MATH	math	A full collection of mathematical calculations
KERAS	keras	An extensive collection of deep learning methods
RANDOM	random	Offers the ability to utilize randomness in data
TENSORFLOW	tensorflow	A comprehensive deep learning library
JSON	json	Allows for the handling of JSON-type files
OPERATING SYSTEM	os	Allows for operating system functions to be used
SYSTEM	sys	Allows for system functions to be used
SQLITE	sqlite3	Allows users to create and manage SQLite databases
STATISTICS	statistics	A full collection of statistical calculations

Figure 2.7 – A table showing some of the most common Python libraries

The libraries included in the previous table are some of the most common ones you will face as you begin your journey in the data science space. In the following section, we will focus on the `math` library to run a few calculations.

General calculations

Within the Python language, we can create variables and assign them specific values, as we previously observed. Following this, we can use the values within these variables to form expressions and conduct mathematical calculations. For example, take the *Arrhenius equation*, commonly used for forecasting molecular stability and calculating the temperature dependence of reaction rates. This equation is commonly used in R&D for two main purposes:

- Optimizing reaction conditions to maximize yields within a synthetic manufacturing process

- Predicting the long-term stability of tablets and pills given changes in temperature and humidity

The equation can be expressed as follows:

$$Arrhenius\ Equation: \quad k = Ae^{-\frac{E_A}{RT}}$$

In this case, k is the rate constant, A is the frequency factor, EA is the activation energy, R is the ideal gas constant, and T is the temperature in **Kelvin** (**K**). We can use this equation to calculate how a change in temperature would affect the rate constant. Let's assume that the current need is to predict what would happen if the temperature changed from 293 K to 303 K. First, we would need to define some variables:

```
>>> from math import exp
>>> EA = 50000
>>> R = 8.31
>>> T1 = 293

>>> exp(-EA / (R*T1))
1.2067e-09
```

We can now reassign the temperature variable another value and recalculate:

```
>>> T2 = 303
>>> exp(-EA / (R*T2))
2.3766e-09
```

In conclusion, this shows that a simple change in temperature nearly doubles the faction!

Lists and dictionaries

Lists and **dictionaries** are two of the most common and fundamental data types in Python. Lists are simply ordered collections of elements (similar to arrays) and can hold elements of the same type, or of different types:

```
>>> homogenousList = ["toluene", "methanol", "ethanol"]
>>> heterogenousList = ["dichloromethane", 3.14, True]
```

The length of any given list can be captured using the len() function:

```
>>> len(heterogenousList)
3
```

List elements can be retrieved using their index location. Remember that all indexes in Python begin at 0, therefore, the first element of this list would be at the 0 index:

```
>>> heterogenousList[0]
dichloromethane
>>> heterogenousList[1]
3.14
```

Unlike their primitive counterparts, lists are mutable, as they can be altered after their creation. We can add another element to a list using the append() function:

```
>>> len(homogenousList)
3
>>> homogenousList.append("acetonitrile")
>>> len(homogenousList)
4
```

Dictionaries, on the other hand, are often used for their association of **keys** and **values**. Given a dictionary, you can specify the name of a *key* along with its corresponding *value*. For example, a chemical inventory list containing chemical names and their expiration dates would not work well within a standard Python list. The chemical names and their dates cannot be *associated* together very easily in this formation.

However, a dictionary would be a perfect way to make this association:

```
>>> singleChemical = {"name" : "acetonitrile",
        "exp_date" : "5/26/2021"}
```

This dictionary now represents the elements of a single *chemical*, in the sense that there is a key assigned to it for a name, and another for an expiration date. You can retrieve a specific value within a dictionary by specifying the key:

```
>>> singleChemical["name"]
    acetonitrile
```

To construct a full inventory of chemicals, you would need to create multiple dictionaries, one for each chemical, and add them all to a single list. This format is known as **JSON**, which we will explore in more detail later in this chapter.

Arrays

Arrays in Python are analogous to lists in the sense that they can contain elements of different types, they can have multiple duplicates, and they can be changed and mutated over time. Arrays can be easily extended, appended, cleared, copied, counted, indexed, reversed, or sorted using simple functions. Take the following as an example:

1. Let's go ahead and create an array using numpy:

    ```
    import numpy as np
    newArray = np.array([1,2,3,4,5,6,7,8,9,10])
    ```

2. You can add another element to the end of the list using the append() function:

    ```
    >>> newArray= np.append(newArray,25)
    >>> newArray
      [1,2,3,4,5,6,7,8,9,10,25]
    ```

3. The length of the array can be determined using the `len()` function:

```
>>> len(newArray)
11
```

4. The array can also be sliced using brackets and assigned to new variables. For example, the following code takes the first five elements of the list:

```
>>> firstHalf = newArray[:5]
>>> firstHalf
[1,2,3,4,5]
```

Now that we have mastered some of the basics when it comes to Python, let's dive into the more complex topic of *functions*.

Creating functions

A **function** in Python is a way to organize code and isolate processes, allowing you to define an explicit input and an explicit output. Take, for example, a function that squares numbers:

```
def squaring_function(x):
    # A function that squares the input
    return x * x
```

Functions are *first class* in the sense that they can be assigned to variables or subsequently passed to other functions:

```
>>> num = squaring_function(5)
>>> print(num)
25
```

Depending on their purpose, functions can have multiple inputs and outputs. It is generally accepted that a function should serve one specific purpose and nothing more.

Iteration and loops

There will be many instances where a task must be conducted in a repetitive or iterative manner. In the previous example, a single value was squared, however, what if there were 10 values that needed to be squared? You could either manually run this function over and over, or you could iterate it with a **loop**. There are two main types of loops: `for` loops and `while` loops. A `for` loop is generally used when the number of iterations is known. On the other hand, a `while` loop is generally used when a loop needs to be broken based on a given condition. Let's take a look at an example of a `for` loop:

```
input_list = [1, 2, 3, 4, 5, 6, 7, 8, 9, 10]
output_list = []
for val in input_list:
    squared_val = squaring_function(val)
    output_list.append(squared_val)
    print(val, " squared is: ", squared_val)
```

First, a list of values is defined. An empty list – where the squared values will be written to – is then created. We then iterate over the list, square the value, append (add) it to the new list, and then print the value. Although `for` loops are great for iterating, they can be incredibly slow in some cases when handling larger datasets.

However, `while` loops can also be used for various types of iterations, specifically when the iteration is to cease when a condition is met. Let's take a look at an example of a `while` loop:

```
current_val = 0
while current_val < 10:
    print(current_val)
    current_val += 1
```

Now that we have gained a stronger understanding of loops and how they can be used, let's explore a more advanced form of iteration known as *list comprehension*.

List comprehension

Like `for` loops, **list comprehension** allows the iteration of a process using a powerful single line of code. We can replicate the previous example of squaring values using this single line:

```
>>> my_squared_list = [squaring_function(val) for val
in input_list]
```

There are three main reasons why you should use list comprehension:

- It can reduce several lines of code down to a single line, making your code much neater.

- It can be significantly faster than its `for` loop counterparts.

- It is an excellent interview question about writing efficient code. Hint hint.

DataFrames

DataFrames using the `pandas` library are arguably among the most common objects within the Python data science space. DataFrames are analogous to structured tables (think of an **Excel** spreadsheet), allowing users to structure data in rows and columns. DataFrames can be constructed using lists, dictionaries, or even full **CSV** documents from your local computer. A `DataFrame` object can be constructed as follows:

```
>>> import pandas as pd
>>>
 df = pd.DataFrame([[1,2,3],[4,5,6],[7,8,9]],columns =
['col1','col2', 'col3'])
>>> print(df)
```

This will give the following output:

	col1	col2	col3
0	1	2	3
1	4	5	6
2	7	8	9

Figure 2.8 – A table showing the results of the DataFrame object

Almost every parameter within a `DataFrame` object can be changed and altered to fit the data within it. For example, the columns can be relabeled with full words:

```
>>> df.columns = ["ColumnA", "ColumnB", "ColumnC"]
```

New columns can be created representing an output of a mathematical function. For example, a column representing the squared values of `ColumnC` can be prepared:

```
>>> df["ColumnC_Squared"] = df["ColumnC"] ** 2
>>> print(df)
```

The output of this is as follows:

	ColumnA	ColumnB	ColumnC	ColumnC_Squared
0	1	2	3	9
1	4	5	6	36
2	7	8	9	81

Figure 2.9 – A table showing the results of the DataFrame object

Alternatively, DataFrames can be prepared using a pre-existing CSV file from your local machine. This can be accomplished using the `read_csv()` function:

```
>>> import pandas as pd
>>> df = pd.read_csv('dataset_lipophilicity_sd.csv')
```

Instead of importing the entire dataset, a specific set of columns can be selected:

```
>>> df = df[["ID", "TPSA", "MolWt", "LogP"]]
>>> df.head()
```

The output of this is as follows:

	ID	TPSA	MolWt	LogP
0	25239916	74.73	367.405	4.34672
1	25239917	58.36	235.331	1.34040
2	25239918	90.48	463.365	3.50380
3	25239919	64.63	265.290	2.02770
4	25239920	41.93	373.468	3.83100

Figure 2.10 – A table showing the results of the DataFrame object

Alternatively, the `tail()` function can also be used to view the last few rows of data:

```
>>> df.tail()
```

DataFrames are some of the most common forms of data handling and presentation within Python, as they resemble standard 2D tables that most people are familiar with. A more efficient alternative for handling larger quantities of data is to use **Apache Spark DataFrames**, which can be used with the `PySpark` library.

Now that we are able to manage and process data locally on our machines, let's take a look at how we could retrieve data from external sources using *API requests*.

API requests and JSON

In some instances, data will not be available locally on your computer and you will need to retrieve it from a remote location. One of the most common ways to send and receive data is in the form of an **application programming interface (API)**. The main idea behind an API is to use HTTP requests to obtain data, commonly communicated in the **JSON** format. Let's take a look at an example:

```
import requests
r = requests.get('https://raw.githubusercontent.com/alkhalifas/
node-api-books/master/services/books.json')
data = r.json()
```

Think of a JSON as a list of dictionaries in which each dictionary is an element. We can select specific elements in the list based on their index locations. In Python, we begin the count at 0, therefore, the first item in our list of dictionaries will have the index location of 0:

```
>>> data[0]
```

This gives us the following:

```
{
    "_id": "5fe7cf66c88fa116803c008c",
    "title": "Gone with the Wind",
    "author": "Margaret Mitchell",
    "pubDate": "1936-06-30T16:00:00.000Z",
    "edition": 1,
    "type": "HARD_COVER"
},
```

Figure 2.11 – A sample of the results obtained from an HTTP request

The *values* within the dictionary can be accessed using their corresponding *keys*:

```
>>> data[0]["type"]
HARD_COVER
```

In a similar way to CSV files, JSON files can also be imported into DataFrames using the `read_json()` function.

Parsing PDFs

Unlike the many structured forms of data we have imported into Python, such as CSV and JSON files, you will often encounter data in its unstructured form – for example, text files or PDFs. For most applications that use **natural language processing** (**NLP**), data is generally gathered in its unstructured form and then preprocessed into a more structured state. PDF documents are among the most common sources of text-based data and there are numerous libraries available allowing users to parse their contents more easily. Here, we will explore a new library known as `tika` – one of the most popular in the open source community. We can begin by installing the library using `pip`:

```
alkhalifas@titanium ~ % pip install tika
```

We can then go ahead and read the specific PDF file of interest:

```
from tika import parser
raw = parser.from_file("./datasets/COVID19-CDC.pdf")
print(raw['content'])
```

The data within `raw['content']` will be the text of the PDF file parsed by the `tika` library. This data is now ready to be used and preprocessed in a subsequent NLP application.

Pickling files

The vast majority of the documents we have handled so far are files that are commonly saved locally to your computer – for example, PDFs, CSVs, and JSON files. So, how do we save a Python object? If you have an important list of items you wish to save – perhaps the chemicals from our previous example – you will need a way to save those files locally to use at a later time. For this, most data scientists use `pickle`. The `pickle` library allows you to save and store Python objects for use at a later time in the form of `.pkl` files. These files can later be imported back to Python and used for new tasks. This is a process known as *serializing* and *deserializing* objects in Python. Let's take a look at an example of using a `.pkl` file. We start by importing the `pickle` library and then creating a list of items:

```
>>> import pickle
>>> cell_lines = ["COS", "MDCK", "L6", "HeLa", "H1", "H9"]
```

In order to save the list as a `.pkl` file, we will need to specify a location for the file to be saved in. Note that we will be using the wb mode (that is, the *write binary* mode). We will then use the `dump()` function to save the contents:

```
>>> pickledList = open('./tmp/cellLineList.pkl', 'wb')
>>> pickle.dump(cell_lines, pickledList)
```

Whether the file was saved locally or shared with a colleague, it can subsequently be loaded back into Python using the `load()` command in a similar manner:

```
>>> pickledList = open('./tmp/cellLineList.pkl', 'rb')
>>> cell_lines_loaded = pickle.load(pickledList)
>>> print(cell_lines_loaded)
["COS", "MDCK", "L6", "HeLa", "H1", "H9"]
```

Notice that in the preceding example, we switched between two arguments – wb (write binary) and rb (read binary) – depending on the task. These are two modes that can be selected to load and save files. There are a number of other options that can be used. The main distinction that should be noted here is the use of *binary* formats. On Windows, opening the files in binary mode will address end-of-line characters in text files mostly seen in ASCII files. The following table provides an overview of some of the most common modes:

MODE	ABBREVIATION	USED FOR
READ	r	Open the file to view only
WRITE	w	Open the file to view and edit
APPEND	a	Append a line to the file
READ BINARY	rb	Read the file as a binary file
WRITE BINARY	wb	Write to the file as a binary file

Figure 2.12 – A table showing the most common read/write modes

Now that we have gained a basic understanding of APIs and the operations we can perform on data, let's take a look at the use of **object-oriented programming** (**OOP**) as it relates to Python.

Object-oriented programming

Similar to many other languages – such as C++, Java, and C# – the concept of OOP can also be used in Python. In OOP, the main aim is to use **classes** to organize and **encapsulate** data objects and their associated functions.

Let's explore an example of OOP in the context of chemical inventory management. Most modern biotechnology companies have extensive inventory systems that monitor their stock of in-house chemicals. Inventory systems allow companies to ensure that supplies do not run out and expiration dates are adequately monitored, along with many other tasks. Now, we will use our current knowledge of Python alongside the concept of OOP to construct an inventory management system:

1. We begin by importing two libraries we will need to manage our dates:

    ```
    import datetime
    from dateutil import parser
    ```

2. Next, we define a name for the class using the following syntax:

    ```
    class Chemical:
    ```

3. We then construct a portion of code known as the constructor:

    ```
    class Chemical:
            def __init__(self, name, symbol, exp_date,
    count):
                self.name = name
                self.symbol = symbol
                self.exp_date = exp_date
                self.count = count
    ```

    The purpose of the **constructor** is to act as a general blueprint of the objects we intend to create. In our case, the object will be a chemical object, and each chemical object will consist of a number of parameters. In this case, the chemical has a *name*, a *symbol*, an *expiration date*, and a *count*. We use the __init__ function to initialize or create this object, and we use self to refer to that specific instance of a class. For example, if we created two chemical objects, self.name could be acetonitrile for one object and methanol for another.

4. Next, we can define a few functions that are relevant to our class. These functions are class-specific and are *tied* to the class in such a way that they can only be accessed through it. We call these functions **member functions**. Within each function, we add `self` as an argument to tie the function to the specific instance of interest. In the following example, we will create an `isExpired()` function that will read the expiration date of the chemical and return a `True` value if expired. We begin by determining today's date, and then retrieve the object's date using the `self.exp_date` argument. We then return a Boolean value that is the product of a comparison between the two dates:

```
def isExpired(self):
    todays_date = datetime.datetime.today()
    exp_date = parser.parse(self.exp_date)
    return todays_date > exp_date
```

5. We can test this out by creating a new object using the `Chemical` class:

```
>>> chem1 = Chemical(name="Toluene", symbol="TOL", exp_
date="2019-05-20", count = 5)
```

6. With that, we have constructed a chemical object we call `chem1`. We can retrieve the fields or attributes of `chem1` by specifying the name of the field:

```
>>> Print(chem1.name)
Toluene
```

Notice that the `name` field was not followed by parentheses in the way we have previously seen. This is because `name` is only a field that is associated with the class and not a function.

7. We can use our function by specifying in a similar manner:

```
>>> print(chem1.isExpired())
True
```

Notice that the function was followed by parentheses, however, no argument was added within it, despite the fact that the function itself within the class specifically contained the `self` argument.

8. We can create many instances of our class, with each instance containing a different date:

```
>>> chem2 = Chemical(name="Toluene", symbol="TOL", exp_
date="2021-11-25", count = 4)
>>> chem3 = Chemical(name="Dichloromethane",
symbol="DCM", exp_date="2020-05-13", count = 12)
>>> chem4 = Chemical(name="Methanol", symbol="MET", exp_
date="2021-01-13", count = 5)
```

Each of these objects will return their respective values when followed by either their fields or functions.

9. We can also create functions to summarize the data within a particular instance of an object:

```
def summarizer(self):
        print("The chemical", self.name, "with the
symbol (",self.symbol,") has the expiration date", self.
exp_date)
```

10. We can then call the `summarizer()` function on any of the chemical objects we have created in order to retrieve a human-readable summary of its status:

```
>>> print(chem1.summarizer())
The chemical Toluene with the symbol ( TOL ) has the
expiration date 2019-05-20
```

11. The functions we have written so far have not taken any additional arguments and have simply retrieved data for us. Chemical inventory systems often need to be updated to reflect items that have expired or been consumed, thereby altering the count. Functions can also be used to change or alter the data within an object:

```
def setCount(self, value):
        self.count = value
```

12. We can simply add `value` as an argument to set that instance's count (represented by `self.count`) to the corresponding value. We can test this using one of our objects:

```
>>> chem1 = Chemical(name="Toluene", symbol="TOL", exp_
date="2019-05-20", count = 5)
>>> chem1.count
  5
>>> chem1.setCount(25)
```

```
>>> chem1.count
25
```

There are many other uses, applications, and patterns for OOP that go above and beyond the example we have just seen. For example, inventory systems would not only need to maintain their inventories, but they would also need to manage the expiry dates of each item, record sales details, and have methods and functions to compile and report these metrics. If you are interested in the development of classes within Python, please visit the official Python documentation to learn more (`https://docs.python.org/3/tutorial/classes.html`).

Tutorial – working with Rdkit and BioPython

In the previous tutorial, we saw various examples of how Python can be used to calculate properties, organize data, parse files, and much more. In addition to the libraries we have worked with thus far, there are two others in particular we need to pay close attention to when operating in the fields of Biotechnology and Life Sciences: **Rdkit**, and **BioPython**. In the following sections, we will look at a few examples of the many capabilities available in these packages. With this in mind, let us go ahead and get started!

Working with Small Molecules and Rdkit

One of the most common packages data scientists use when handling data relating to small molecules is known as **rdkit**. Rdkit is an open-source cheminformatics and machine learning package with numerous useful functionalities for both predictive and generative purposes. The `rdkit` package includes many different tools and capabilities to the point that we would need a completely second book to cover in total. Highlighted below are five of the most common applications this package is generally known for, as seen in *Figure 2.13*:

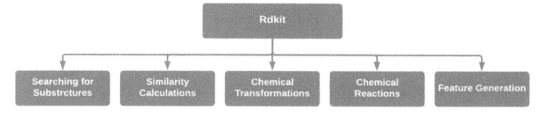

Figure 2.13 – Some of the main functionality in the rdkit package

Let us go ahead and example a few of these capabilities in order to introduce ourselves to the `rdkit` package.

Working with SMILES Representations

Similar to some of the packages we have seen already, rdkit is organized by classes. Let us now take advantage of the Chem class to load a SMILES representation in a few simple steps.

We will start off by importing the Chem class:

```
from rdkit import Chem
```

One of the easiest and most common ways to transfer 2D molecular structures from one Python script to another is by using a SMILES representation. For example, we can describe a **Quaternary Ammonium Compound** (**QAC**) using its SMILES representation as seen here:

```
SMILES = "[Br-].[Br-].CCCCCCCCCC[N+]1=CC=C(CCCC2=CC=[N+]
(CCCCCCCCCC)C=C2)C=C1"
```

We can use the MolFromSmiles function within the Chem class to load our SMILES representation into rdkit:

```
molecule = Chem.MolFromSmiles(SMILES)
molecule
```

Upon printing the molecule variable we assigned above, a figure of the molecule will be returned as shown in *Figure 2.14*:

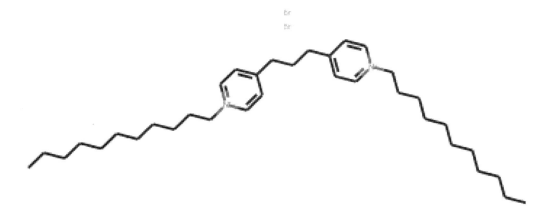

Figure 2.14 – A 2D representation of the QAC using rdkit

Notice that we did not need any additional packages to print this figure as rdkit is quire comprehensive and has everything you need to run these visualizations. In the following section, we will see another example of rdkit as it comes to similarity calculations.

With the structure now loaded, there are many different applications and calculations one can do. One of the most common methods to do here is searching for substructures within the molecule. We can accomplish this by using the MolFromSmarts function within rdkit:

```
tail_pattern = Chem.MolFromSmarts('CCCCCCCCCC')
patter
```

Upon executing this, we yield the following figure showing the substructure of interest:

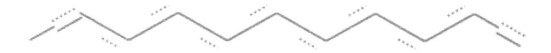

Figure 2.15 – A 2D substructure of interest

With the main molecule of interest and the pattern now both loaded, we can use the HasSubstructMatch function to determine if the substructure exists:

```
molecule.HasSubstructMatch(tail_pattern)
```

Upon executing this code, the value of True will be returned indicating that the structure does exist. On the other hand, if another substructure such as a phenol were to be run, the value would return as False, since that substructure does not exist in the main molecule.

In addition, similarity calculations can be run using the DataStructs class within rdkit. We can being by importing the class, and entering two molecules of interest:

```
from rdkit import DataStructs
from rdkit.Chem import Draw
mol_sim = [Chem.MolFromSmiles('[Br-].[Br-].
CCCCCC[N+]1=CC=C(CCCC2=CC=[N+](CCCCC)C=C2)C=C1'), Chem.
MolFromSmiles('[Br-].[Br-].CCCCCCCCCC[N+]1=CC=C(CCCC2=CC=[N+]
(CCCCCCCCCC)C=C2)C=C1')]
```

If we compared the two molecules visually using the `MolFromSmiles` method we saw previously, we can see that there exists a minor difference between the two structures, being the double bonds in the hydrophobic tails in one of the molecules.

Next, we can use the `RDKFingerprint` function to calculate the fingerprints:

```
fps = [Chem.RDKFingerprint(x) for x in mol_sim]
```

Finally, we can use `CosineSimilarity` metric to calculate how different the two structures are:

```
DataStructs.FingerprintSimilarity(fps[0],fps[1],
metric=DataStructs.CosineSimilarity)
```

This calculation will yield a value of approximately 99.14%, indicating that the structures are mostly the same with the exception of a minor difference.

Summary

Python is a powerful language that will serve you well, regardless of your area of expertise. In this chapter, we discussed some of the most important concepts when working with the command line, such as creating directories, installing packages, and creating and editing Python scripts. We also discussed the Python programming language quite extensively. We reviewed some of the most commonly used IDEs, general data types, and calculations. We also reviewed some of the more complex data types such as lists, DataFrames, and JSON files. We also looked over the basics of APIs and making HTTP requests, and we introduced OOP with regard to Python classes. All of the examples we explored in this chapter relate to applications commonly discussed within the field of data science, so having a strong understanding of them will be very beneficial.

Although this chapter was designed to introduce you to some of the most important concepts in data science (such as variables, lists, JSON files, and dictionaries), we were not able to cover them all. There are many other topics, such as tuples, sets, counters, sorting, regular expressions, and many facets of OOP that we have not discussed. The documentation for Python – both in print and online – is extensive and mostly free. I would urge you to take advantage of these resources and learn as much as you can from them.

Within this chapter, we discussed many ways to handle small amounts of data when it comes to slicing it around and running basic calculations. At the enterprise level, data generally comes in significantly larger quantities, and therefore, we'll need the right tool to tackle it. That tool is **Structured Query Language** (**SQL**), and we will get acquainted with this in the next chapter.

3
Getting Started with SQL and Relational Databases

According to a recent article published in the *Journal of Big Data Analytics and Its Applications*, every 60 seconds on the internet, the following happens:

- 700,000 status updates are made.

- 11 million messages are sent.

- 170 million emails are received.

- 1,820 **terabytes** (**TB**) of new data is created.

It would be an understatement to claim that data within the business landscape is growing rapidly at an unprecedented rate. With this major explosion of information, companies around the globe are investing a great deal of capital in an effort to effectively capture, analyze, and deliver benefits from this data for the company. One of the main methods by which data can be managed and subsequently retrieved to provide actionable insights is through **Structured Query Language** (**SQL**).

Similar to how we used the **Terminal command line** to create **directories**, or Python to run calculations, you can use **SQL** to create and manage databases either locally on your computer or remotely in the cloud. SQL comes in many forms and flavors depending on the platform you choose to use, each of which contains a slightly different syntax. However, SQL generally consists of four main types of language statements, and these are outlined here:

- **Data Manipulation Language** (**DML**): Querying and editing data
- **Data Definition Language** (**DDL**): Querying and editing database tables
- **Data Control Language** (**DCL**): Creating roles and adding permissions
- **Transaction Control Language** (**TCL**): Managing database transactions

Most companies around the globe have their own separate *best practices* when it comes to DDL, DCL, and TCL and how databases are integrated into their enterprise systems. However, DML is generally the same and is often the main focus of any given data scientist. For these purposes, we will focus this chapter on applications relating to DML. By the end of this chapter, you will have gained a strong introduction to some of the most important database concepts, fully installed MySQL Workbench on your local machine, and deployed a full **Amazon Web Services Relational Database Service** (**AWS RDS**) server to host and serve your data. Note that all of these capabilities can later be recycled for your own endeavors. Let's get started!

We will cover the following topics in this chapter:

- Exploring relational databases
- Tutorial: Getting started with MySQL

Technical requirements

In this chapter, we will explore some of the main concepts behind relational databases, their benefits, and their applications. We will focus on one specific *flavor* of a relational database known as MySQL. We will use **MySQL** through its common **User Interface** (**UI**), known as **MySQL Workbench**. Whether you are using a Mac or a PC, the installation process for MySQL Workbench will be very similar, and we will walk through this together later in this chapter. This interface will allow you to interact with a database that can either be hosted on your local machine or far away in the cloud. Within this chapter, we will deploy and host our database server in the AWS cloud, and you will therefore need to have an AWS account. You can create an account by visiting the AWS website (https://aws.amazon.com/) and signing up as a new user.

Exploring relational databases

There are numerous types of databases—such as **Object-Oriented (OO)** databases, graph databases, and relational databases—each of which offers a particular capability.

OO databases are best used in conjunction with OO data. Data within these databases tends to consist of objects that contain members such as fields, functions, and properties; however, relations between objects are not well captured.

On the other hand, **graph databases**, as the name suggests, are best used with data consisting of nodes and edges. One of the most common applications for graph databases in the biotechnology sector is in small-molecule drug design. Molecules consist of nodes and edges that connect together in one form or another—these relations are best captured in graph databases. However, the relationships between molecules are not well captured here either.

Finally, **relational databases**, as the name suggests, are best used for databases in which relations are of major importance. One of the most important applications of relational databases in the biotechnology and healthcare sectors surrounds patient data. Relational databases are used with patient data due to its complex nature. Patients will have many different fields, such as name, address, location, age, and gender. A patient will be prescribed a number of medications, each medication containing its own sub-fields, such as a lot of numbers, quantities, and expiration dates. Each lot will have a manufacturing location, manufacturing date, and its associated ingredients, and each ingredient will have its own respective parameters, and so on. The complex nature of this data and its associated relations is best captured in a relational database.

Relational databases are standard digital databases that host tables in the form of columns and rows containing relations to one another. The relationship between two tables exists in the form of a **Unique Identifier (UID)** or **Primary Key (PK)**. This key acts as a unique value for each of the rows within a single table, allowing users to match rows of one table to their respective rows in another table. The tables are not technically connected in any way; they are simply referenced or related to one another. Let's take a look at a simple example when it comes to patient data.

If we were to create a table of patient data containing patients' names, contact information, their pharmacies, and prescribing physicians, we would likely come up with a table similar to this:

Patient ID	Name	Address	Phone	PCP Name	PCP Address	PCP Phone	Pharmacy Name	Pharmacy Address	Pharmacy Phone
1	John Doe	123 Joes St	617-111-1234	Dr. Howard	121 List St	617-123-4567	Health Pharmacy	151 Main St	617-222-1234
2	Jane Doe	234 Ann St	617-222-2345	Dr. Fine	131 String St	617-123-4567	Health Pharmacy	151 Main St	617-222-1234
3	Joseph Doe	345 Lore St	617-333-3456	Dr. Howard	121 List St	617-123-4567	Daily Pharmacy	181 Git St	617-222-3456

Figure 3.1 – Table showing an example patient dataset

From an initial perspective, this table makes perfect sense. We have the patient's information listed nicely on the left, showing their name, address, and phone number. We also have the patient's respective **Primary Care Physician** (**PCP**), showing their associated names, addresses, and phone numbers. Finally, we also have the patient's respective pharmacy and their associated contact information. If we were trying to generate a dataset on all of our patients and their respective PCPs and pharmacies, this would be the perfect table to use. However, storing this data in a database would be a different story.

Notice that some of the PCP names and their contact information are repeated. Similarly, one of the pharmacies appears more than once in the sense that we have listed the name, address, and phone number twice in the same table. From the perspective of a three-row table, the repetition is negligible. However, as we scale this table from 3 patients to 30,000 patients, the repetition can be very costly from both a database perspective (to host the data) and a computational perspective (to retrieve the data).

Instead of having one single table host all our data, we can have multiple tables that when joined together temporarily for a particular purpose (such as generating a dataset) would be significantly less costly in the long run. This idea of splitting or **normalizing** data into smaller tables is the essence of relational databases. The main purpose of a **relational database** is to provide a convenient and efficient process to store and retrieve information while minimizing duplication as much as possible. To make this database more "relational", we can split the data into three tables—pharmacies, patients, and PCPs—so that we only store each entry once but use a system of keys to reference them.

Here, you can see a **Unified Modeling Language** (UML) diagram commonly used to describe relational databases. The connection between the tables indicated by the single line splitting into three others is a way to show a *one-to-many* relationship. In the following case, one pharmacy can have many patients and one PCP can have many patients, but a patient can have only one PCP and only one pharmacy. The ability to quickly understand a database design and translate that to a UML diagram (or vice versa) is an excellent skill to have when working with databases and is often regarded as an excellent interview topic—I actually received a question on this a few years ago:

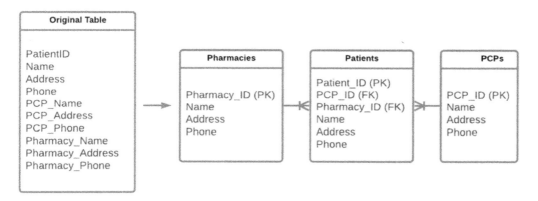

Figure 3.2 – The process of normalizing a larger table into smaller relational tables

The previous diagram provides a representation of how the original table can be split up. Notice that each of the tables has an ID; this ID is its UID or PK. Tables that are connected to others are referenced using a **Foreign Key** (**FK**). For example, a PCP can have multiple patients, but each patient will only have one PCP; therefore, each patient entry will need to have the FK of its associated PCP. Pretty interesting, huh?

The process of separating data in this fashion is known as database **normalization**, and there are a number of rules most relational databases must adhere to in order to be normalized properly. Data scientists do not often design major enterprise databases (we leave these tasks to database administrators). However, we often design smaller **proof-of-concept** databases that adhere to very similar standards. More often, we interact with significantly larger enterprise databases that are generally in a relational state or in the form of a data lake. In either case, a strong foundational knowledge of the structure and general idea behind relational databases is invaluable to any data scientist.

Database normalization

There are a number of rules we must consider when preparing a database that is **normalized**, often referred to as **normal forms**. There are three normal forms that we will briefly discuss within the context of relational databases, as follows:

First normal form

In order to satisfy the **First Normal Form** (**1NF**) of database normalization, the values in each of the cells must be **atomic** in the sense that each cell contains only one type of data. For example, a column containing addresses such as **5 First Street, Boston MA 02215** contains non-atomic data and violates this rule as it contains street numbers, street names, cities, states, and zip codes in one cell. We can normalize this data by splitting it into five columns, as follows:

Primary Key	Street Number	Street Name	City	State	Zip code
1	5	First Street	Boston	MA	02215
2	451	Second Street	Boston	MA	02215
3	461	Third Street	Boston	MA	02215

Figure 3.3 – Table showing how addresses can be split into individual atomic cells

Now that we have gained a better understanding of the first form of data normalization, let's go ahead and explore the second form.

Second normal form

In order to satisfy the conditions of the **Second Normal Form** (**2NF**), the table must have a **PK** acting as a UID, and that all the fields excluding the PK must be functionally dependent on the entire key. For example, we can have a table with a list of all patients in the sense that the first name, last name, and phone number of the patients are dependent on the PK. This key represents the patient and nothing else. We can also have another table representing the PCPs, their locations, associated hospitals, and so on. However, it would not be appropriate from the perspective of database normalization to add this information to a table representing a patient.

To satisfy this condition with the patient database, we would need to split the table such that the patient data is in one table and the associated PCP is in a second table, connected via an FK, as we saw in the earlier example.

Third normal form

In order to satisfy the conditions for the **Third Normal Form** (**3NF**), we must satisfy the conditions of the 1NF and 2NF, in addition to ensuring that all the fields within the table are functionally independent of each other—in other words, no fields can be calculated fields. For example, a field titled *Age* with values of 27 years, 32 years, and 65 years of age would not be appropriate here as these are calculated quantities. Instead, a field titled *Date of Birth* with the associated dates could be used to satisfy this condition.

Most data scientists spend little time structuring major databases and normalizing them; however, a great deal of time is spent understanding database structures and forming queries to retrieve data correctly and efficiently. Therefore, a strong foundational understanding of databases will always be useful.

Although database administrators and data engineers tend to spend more time on structuring and normalizing databases, a great deal of time is spent at a data scientist's end when it comes to understanding these structures and developing effective queries to retrieve data correctly and efficiently. Therefore, a strong foundational understanding of databases will always be useful regardless of the type of database being used.

Types of relational databases

There are several different *flavors* of **SQL** that you will likely encounter depending on the commercial database provider. Many of these databases can be split into two general categories: **open source** or **enterprise**. Open source databases are generally free, allowing students, educators, and independent users to use their software without restrictions, depending on their specific terms and conditions. Enterprise databases, on the other hand, are commonly seen in large companies. We'll now look at some of the most common databases found in most industries, including tech, biotech, and healthcare.

Open source

The following list shows some common open source databases:

- **SQLite**: An **Atomicity, Consistency, Isolation, Durability** (**ACID**)-compliant **relational database management system** (**RDBMS**) commonly used in smaller locally hosted projects. SQLite can be used within the Python language to store data.

- **MySQL**: Although not ACID-compliant, MySQL provides similar functionality to SQLite but at a larger scale, allowing for greater amounts of data to be stored, and providing multiuser access.

- **PostgreSQL**: An ACID-compliant database system that provides faster data processing and is better suited for databases with larger user bases.

Let's now explore some enterprise options instead.

Enterprise

The following list shows some common enterprise databases:

- **AWS RDS**: A cloud-based relational database service that provides scalable and cost-efficient services to store, manage, and retrieve data.
- **Microsoft SQL Server**: An enterprise RDMS similar to that of MySQL Workbench with cloud-hosted services to store and retrieve data.
- **Systems Applications and Products in data processing** (**SAP**): An RDBS solution for storing and retrieving services, commonly used for inventory and manufacturing data.

Now, let's get hands-on with MySQL.

Tutorial – getting started with MySQL

In the following tutorial, we will explore one of the most common processes to launch a cloud-based server to host a private relational database. First, we will install an **instance** of MySQL—one of the most popular database management platforms. We will then create a full free-tier **AWS RDS server** and connect it to the MySQL instance. Finally, we will upload a local **Comma-Separated Values** (**CSV**) file pertaining to small-molecule toxicity and their associated properties and begin exploring and learning to query our data from our dataset.

You can see a representation of AWS RDS being connected to the MySQL instance here:

Figure 3.4 – Diagram showing that MySQL will connect to AWS RDS

> **Important note**
>
> Note that while this tutorial involves the creation of a database within this AWS RDS instance for the toxicity dataset, you will be able to recycle all the components for future projects and create multiple new databases without having to repeat the tutorial. Let's get started!

Installing MySQL Workbench

Of the many database design tools available, MySQL Workbench tends to be the easiest to design, implement, and use. MySQL Workbench is simply a **Graphical UI (GUI)** virtual database design tool developed by Oracle and allows users to create, design, manage, and interact with databases for various projects. Alternatively, MySQL can be used in the form of MySQL Shell, allowing users to interact with databases using the terminal command line. For those interested in working from the terminal command line, MySQL Shell can be downloaded by navigating to `https://dev.mysql.com/downloads/shell/` and using the **Microsoft Installer (MSI)**. However, for the purposes of this tutorial, we will be using MySQL Workbench instead for its user-friendly interface. Let's get started! Proceed as follows:

1. Getting MySQL Workbench installed on your local computer is fairly simple and easy to complete. Go ahead and navigate to `https://dev.mysql.com/downloads/workbench/`, select your respective operating system, and click **Download**, as illustrated in the following screenshot:

Figure 3.5 – MySQL Installer page

2. Upon downloading the file, click **Install**. Follow through the installation steps if you have specific criteria that need to be met; otherwise, select all the default options. Be sure to allow MySQL to select the standard destination folder followed by the **Complete** setup type, as illustrated in the following screenshot:

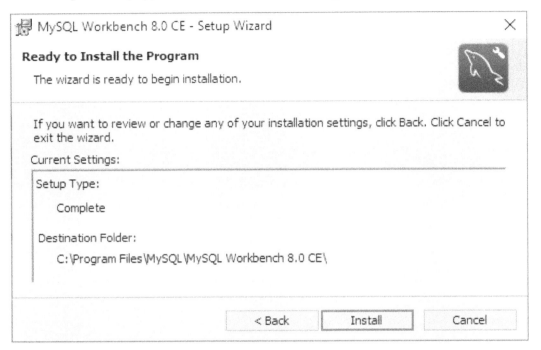

Figure 3.6 – MySQL Installer page (continued)

With that, MySQL Workbench has now been successfully installed on your local machine. We will now head to the AWS website to create a remote instance of a database for us to use. Please note that we will assume that you have already created an AWS account.

Creating a MySQL instance on AWS

Let's create a MySQL instance, as follows:

1. Navigate to `https://www.aws.amazon.com` and log in to your AWS account.
    Once logged in, head to the AWS Management Console, and select **RDS** from the
    Database section, as illustrated in the following screenshot:

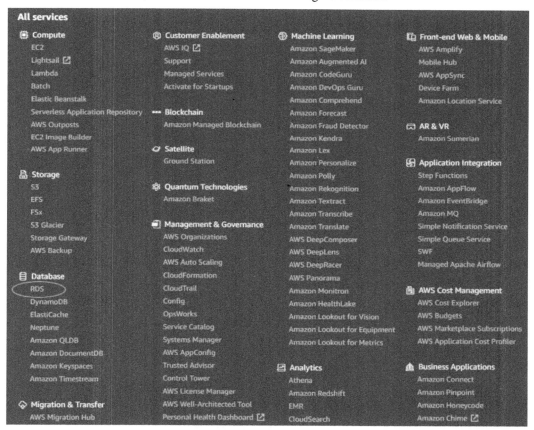

Figure 3.7 – AWS Management Console page

2. From the top of the page, click the **Create Database** button. Select the **Standard create** option for the database creation method, and then select **MySQL** as the engine type, as illustrated in the following screenshot:

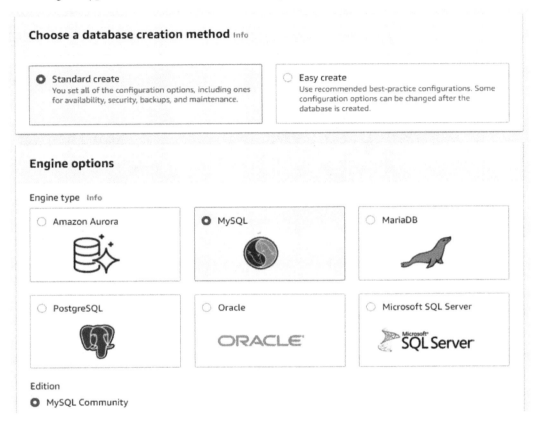

Figure 3.8 – RDS engine options

3. In the **Templates** section, you will have three different options: **Production**, **Dev/ Test**, and **Free tier**. While you can certainly select the first two options if you are planning to use a production-level server, I would recommend selecting the third option, **Free tier**, in order to take advantage of it being free. The following screenshot shows this option being selected:

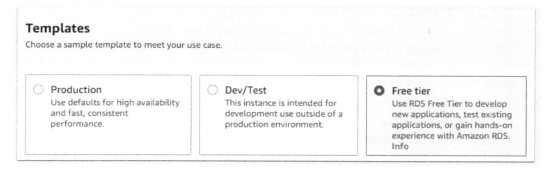

Figure 3.9 – RDS template options

4. Under the **Settings** section, give the name of `toxicitydatabase` for the **DB instance identifier** field. Next, add the master username of `admin`, followed by a password of your choice. You may also take advantage of the **Auto generate a password** feature that AWS provides. If you select this option, the password will be made available to you after the instance is created. The process is illustrated in the following screenshot:

Settings

DB instance identifier Info

Type a name for your DB instance. The name must be unique across all DB instances owned by your AWS account in the current AWS Region.

```
toxicity_database
```

The DB instance identifier is case-insensitive, but is stored as all lowercase (as in "mydbinstance"). Constraints: 1 to 60 alphanumeric characters or hyphens. First character must be a letter. Can't contain two consecutive hyphens. Can't end with a hyphen.

▼ **Credentials Settings**

Master username Info

Type a login ID for the master user of your DB instance.

```
admin
```

1 to 16 alphanumeric characters. First character must be a letter

☑ Auto generate a password

Amazon RDS can generate a password for you, or you can specify your own password

Figure 3.10 – RDS settings options

5. Under the **DB Instance Class** section, select the **Burstable Classes** option followed by a db.t2.micro instance type. For storage, select the default parameters in which the **Storage Type is General Purpose (SSD)** option is selected, and allocate a size of 20 **gigabytes** (**GB**) for the server. Be sure to disable the autoscaling capability as we will not require this feature.

6. Finally, when it comes to connectivity, select your default **Virtual Private Cloud** (**VPC**), followed by the default subnet. Be sure to change the **Public Access** setting to **Yes** as this will allow us to connect to the instance from our local MySQL installation. Next, ensure that the **Password and IAM database authentication** option in the **Database authentication** section is selected, as illustrated in the following screenshot. Next, click **Create Database**:

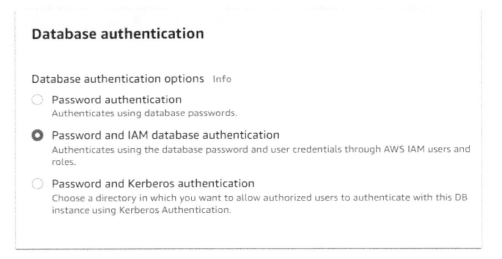

Figure 3.11 – RDS password-generation options

7. Once the database creation process is started, you will be redirected to your RDS console consisting of a list of databases where you will see a toxicitydataset database being created, as illustrated in the following screenshot. Note that the **Status** column of the database will show as **Pending** for a few moments. In the meantime, if you requested that AWS automatically generate a password for your database, you will find that in the **View connection details** button at the top of the page. Note that for security reasons, *these credentials will never be revealed to you again.* Be sure to open these details and copy all of the contents to a safe location. Connecting to this remote database via the local MySQL interface will require the master **Username**, the master **Password**, and the specified **Endpoint** values. With that, we have now created an AWS RDS server and we can now leave AWS in its current state and divert our full attention to MySQL Workbench:

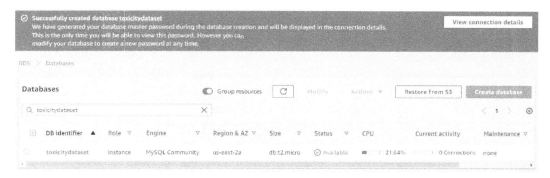

Figure 3.12 – RDS menu

With our AWS infrastructure prepared, let's go ahead and start working with our newly created database.

Working with MySQL

Once the setup on AWS is complete, go ahead and open MySQL Workbench. Note that you may be prompted to restart your computer. Follow these next steps:

1. You should be greeted with a welcome message, followed by various options. To the right of the **MySQL Connections** section, click the + sign to add a new connection, as illustrated in the following screenshot:

MySQL Connections ⊕ ⊗

Figure 3.13 – MySQL Connections button

2. In this menu, we will create a new database connection called `toxicity_db_tutorial`. We will select the `Standard (TCP/IP)` connection method. Change the **Hostname** field to the endpoint that was provided to you in the connection details page in AWS. Next, add the username and password that was either specified or generated for you in AWS. Be sure to store your password in the vault to access it at a later time. Finally, click on the **Test Connection** button. The process is illustrated in the following screenshot. If all steps were correctly followed, you should receive a **Successful connection** response:

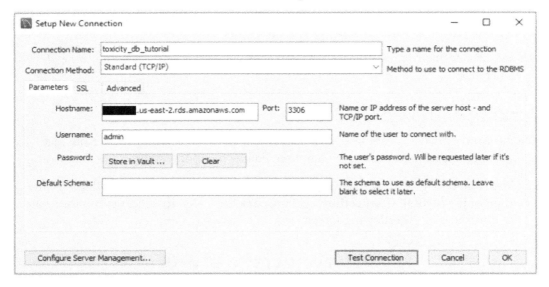

Figure 3.14 – MySQL Setup New Connection menu

3. In the main menu, you should see the new connection listed under the **MySQL Connections** section. Double-click the newly created connection, enter your root computer password (if not saved in the vault), and click **OK**. With that, we have created a new database connection and connected to it using MySQL Workbench.

4. The documentation for MySQL is quite extensive, as it contains a great deal of functionality that almost requires its very own book to fully cover. For the purposes of this tutorial, we will focus our efforts on a subset of functionality that is most commonly used in the data science field. There are three main sections within the **MySQL Workbench** window a user should be aware of—the **schema navigator**, the **query editor**, and the **output window**, as illustrated in the following screenshot:

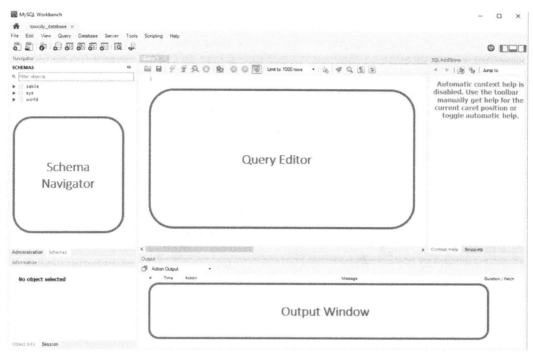

Figure 3.15 – MySQL Workbench preview

The **schema navigator** is the section of the window that allows users to navigate between databases. Depending on who you ask and in what context, the words *schema* and *database* are sometimes synonymous and used interchangeably. Within the context of this book, we will define *schema* as the blueprint of a database, and *database* as the database itself.

The **query editor** is the section of the page where you, as the user, will execute your SQL scripts. Within this section, data can be created, updated, queried, or deleted. You can use the **output window**, as the name suggests, for displaying the output of your executed queries. If a query goes wrong or if something unexpected happens, you will likely find an important message about it listed here.

Creating databases

Before we can begin making any queries, we will need some data to work with. Let's take a look at the currently existing databases in our new server. In the query editor, type in SHOW DATABASES;, and click on the **Execute** ( ) button. You will be provided a list of databases available on your system. Most of these databases were either created from previous projects or by your system to manage data. Either way, let's avoid using those. Now, follow these next steps:

1. We can create a new database using the following SQL statement:

    ```
    CREATE DATABASE IF NOT EXISTS toxicity_db_tutorial;
    ```

2. In the output window, you should see a message confirming the successful execution of this statement. Let's now go ahead and populate our database with a previously existing CSV file. Select the **Schemas** tab from the schema navigator and refresh the list using the icon with two circular arrows. You will see the newly created database appear in the list, as illustrated here:

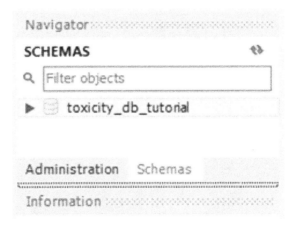

Figure 3.16 – MySQL schema list

3. Next, right-click on the database, and click on **Table Data Import Wizard**. Navigate to and select the dataset_toxicity_sd.csv. CSV file. When prompted to select a destination, select the default parameters, allowing MySQL to create a new table called dataset_toxicity_sd within the toxicity_db_tutorial database. In the **Import Configuration** settings, allow MySQL to select the default datatypes for the dataset. Continue through the wizard until the import process is complete. Given the remote nature of our server, it may take a few moments to transfer the file.

4. Once the file is fully imported into AWS RDS, we are now ready to examine our data and start running a few SQL statements. If you click on the table within the schema navigator, you will see a list of all columns that were imported from the CSV file, as illustrated in the following screenshot:

Table: dataset_toxicity_sd

Columns:

ID	int
smiles	text
toxic	int
FormalCharge	int
TPSA	double
MolWt	double
HeavyAtoms	int
NHOH	int
HAcceptors	int
HDonors	int
Heteroatoms	int
AromaticRings	int
SaturatedRings	int
AromaticOH	int
AromaticN	int
LogP	double

Figure 3.17 – List of columns in the toxicity dataset

Taking a closer look at these columns, we notice that we begin with an ID column that is operating as our PK or UID, with the datatype int, for integer. We then notice a column called smiles, which is a text or string representing the actual chemical structure of the molecule. Next, we have a column called toxic, which is the *toxicity* of the compound represented as *1* for toxic, or *0* for non-toxic. We will call the toxic column our **label**. The rest of the columns ranging from FormalCharge to LogP are the molecules' **attributes** or **features**. One of the main objectives that we will embark upon in a later chapter is developing a predictive model to use these features as input data and attempt to predict the toxicity. For now, we will be using this dataset to explore SQL and its most common **clauses** and **statements**.

Querying data

With the file fully imported into AWS RDS, we are now ready to run a few commands.

We will begin with a simple SELECT statement in which we will retrieve all of our data from the newly created table. We can use the * argument after the SELECT command to denote *all data* within a particular table. The table itself can be specified at the end using the following syntax: <database_name>.<table_name>:.

In the actual command, it looks like this:

```
SELECT * from toxicity_db_tutorial.dataset_toxicity_sd;
```

This gives us the following output:

Figure 3.18 – MySQL Workbench query preview

We will rarely query *all* columns and *all* rows of our data in any given query. More often than not, we will limit the columns to those of interest. We can accomplish this by substituting the * argument with a list of columns of interest, as follows:

```
SELECT
  ID,
    TPSA,
    MolWt,
    LogP,
    toxic
  from toxicity_db_tutorial.dataset_toxicity_sd;
```

Notice that the columns are separated by a comma, whereas the other arguments within the statement are not. In addition to limiting the number of columns, we can also limit the number of rows using a LIMIT clause followed by the number of rows we want to retrieve, as follows:

```
SELECT
  ID,
     TPSA,
     MolWt,
     LogP,
     toxic
  from toxicity_db_tutorial.dataset_toxicity_sd LIMIT 10;
```

In addition to specifying our columns and rows, we can also apply operations to column values such as addition, subtraction, multiplication, and division. We can also filter our data based on a specific set of conditions. For example, we could run a simple query for the ID and smiles columns for all toxic compounds (toxic=1) in the database, as follows:

```
SELECT
  ID,
    SMILES,
      toxic
  from toxicity_db_tutorial.dataset_toxicity_sd
  WHERE toxic=1;
```

Similarly, we could also find all non-toxic compounds by changing the final line of our statement to WHERE toxic = 0 or to WHERE toxic != 1. We can extend this query with the addition of more conditionals within a WHERE clause.

Conditional querying

Conditional queries can be used to filter data more effectively, depending on the use case at hand. We can use the AND operator to query data relating to a specific toxicity and **molecular weight (mol wt)**. Note that the AND operator would ensure that both conditions would need to be met, as illustrated in the following code snippet:

```
SELECT
  ID,
     SMILES,
```

```
    toxic
  from toxicity_db_tutorial.dataset_toxicity_sd
  WHERE toxic=1 AND MolWt > 500;
```

Alternatively, we can also use the OR operator in which either one of the conditions needs to be met—for example, the preceding query would require that the data has a toxic value of 1, and a mol wt of 500, thus returning 26 rows of data. The use of an OR operator here instead would require that *either* one of the conditions is met, thus returning 318 rows instead. Operators can also be used in combination with one another. For example, what if we wanted to query the IDs, smiles representations, and toxicities of all molecules that have 1 hydrogen acceptor, and either 1, 2, or 3 hydrogen donors? We can fulfill this query with the following statement:

```
SELECT
    ID,
        SMILES,
        toxic
  from toxicity_db_tutorial.dataset_toxicity_sd
  WHERE HAcceptors=1 AND (HDonors = 1 OR HDonors = 2 OR HDonors
= 3);
```

It is important to note that when specifying multiple OR operators for values in consecutive order such as with the three HDonors values, the BETWEEN operator can be used instead to avoid unnecessary repetition.

Grouping data

Another common practice when running queries against a database is grouping data by a certain column. Let's take as an example a situation in which we must retrieve the total number of instances of toxic versus non-toxic compounds within our dataset. We could easily query the instances in which the values of 1 or 0 are present using a WHERE statement. However, what if the number of outcomes was 100 instead of 2? It would not be feasible to run this query 100 times, substituting the value in each iteration. For this type of operation, or any operation in which the grouping of values is of importance, we can use a GROUP BY statement in combination with a COUNT function, as illustrated in the following code snippet:

```
SELECT
    ID,
        SMILES,
        COUNT(*) AS count
```

```
from toxicity_db_tutorial.dataset_toxicity_sd
GROUP BY toxic
```

This is shown in the following screenshot:

Figure 3.19 – MySQL Workbench GROUPBY preview

Now that we have explored how to group our data, let's go ahead and learn how to order it as well.

Ordering data

There will be some instances in which your data will need to be ordered in some fashion, either in an ascending or descending manner. For this, you can use an ORDER BY clause in which the sorting column is listed directly after, followed by ASC for ascending or DESC for descending. This is illustrated in the following code snippet:

```
SELECT
  ID,
    SMILES,
  toxic,
    ROUND(MolWt, 2) AS roundedMolWt
 from toxicity_db_tutorial.dataset_toxicity_sd
  ORDER BY roundedMolWt DESC
```

Joining tables

Often when querying records, tables will have been normalized, as previously discussed, to ensure that their data was properly stored in a relational manner. It will often be the case that you will need to merge or join two tables together to prepare your dataset of interest prior to beginning any type of meaningful **Exploratory Data Analysis (EDA)**. For this type of operation, we can use a JOIN clause to join two tables together.

Using the same **Table Data Import Wizard** method for the previous dataset, go ahead and load the `dataset_orderQuantities_sd.csv` dataset into the same database. Recall that you can specify the same database name but a different table name within the import wizard. Once loaded, we will now have a database consisting of two tables: `dataset_toxicity_sd` and `dataset_orderQuantities_sd`.

If we run a `SELECT *` statement on each of the tables, we notice that the only column the two datasets have in common is the `ID` column. This will act as the UID to connect the two datasets. However, we also notice that the toxicity dataset consists of `1461` rows, whereas the `orderQuantities` dataset consists of 728 rows. This means that one dataset is missing quite a few rows. This is where different types of `JOIN` clauses come in.

Imagine the two datasets as circles to be represented as Venn diagrams with one circle (*A*) consisting of `1461` rows of data, and one circle (*B*) consisting of 728 rows of data. We could join the tables such that we would discard any rows that do not match using an `INNER JOIN` function. Notice that this is represented in the following screenshot as the intersection, or *A* ∩ *B*, of the two circles. Alternatively, we could run a `LEFT JOIN` or `RIGHT JOIN` statement to our data, neglecting the differing contents of one of the datasets represented by *A'* ∩ *B* or *A* ∩ *B'*. Finally, we can use an `OUTER JOIN` statement to join the data, merging the two tables regardless of missing rows, known as a union, or *A* ∪ *B*:

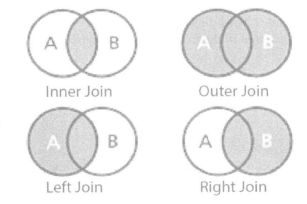

Figure 3.20 – Representation of the four main join methods

As you begin to explore datasets and join tables, you will find the most common JOIN clause is in fact the INNER JOIN statement, which is what we will need for our particular application. We will structure the statement as follows: we will select columns of interest, specify the source table, and then run an inner join in which we match the two ID columns together. The code is illustrated in the following snippet:

```
SELECT
  dataset_toxicity_sd.id,
    dataset_toxicity_sd.SMILES,
    dataset_orderQuantities_sd.quantity_g
FROM
  toxicity_db_tutorial.dataset_toxicity_sd
INNER JOIN toxicity_db_tutorial.dataset_orderQuantities_sd
  ON dataset_toxicity_sd.id = dataset_orderQuantities_sd.id
```

With that, we have successfully managed to take our first dataset (consisting of only the individual research and development molecules and their associated properties) and join it with another table (consisting of the order quantities of those molecules), allowing us to determine which substances we currently have in stock. Now, if we needed to find all compounds with a specific **lipophilicity** (**logP**), we can identify which ones we have in stock and in which quantities.

Summary

SQL is a powerful language when it comes to querying vast amounts of data from relational databases—a skill that will serve you well in all areas of technology and most areas of biotechnology. As most companies begin to grow their database capabilities, you will likely encounter databases of many kinds, especially relational databases.

When it comes to theory, we discussed some of the most important characteristics of relational databases and how data is generally normalized. We looked over an example of patient data and how a table could be normalized to reduce repetition when being stored. We also looked over some of the most common open source and enterprise databases available and readily used on the market today.

When it comes to applications, we put together a robust AWS RDS database server and deployed it to the cloud. We then connected our local instance of MySQL to that server and populated it with a new database using a CSV file. We then went over some of the most common SQL statements and clauses used in the industry today. We looked at ways to select, filter, group, and order data. We then looked at an example of joining two tables together and understood the different join methods available to us.

Although this book was designed to introduce you to some of the most important core concepts every data scientist should know, there were many other topics within SQL that we did not cover. I would urge you to review the MySQL documentation to learn about the many other exciting statements available, allowing you to query data in many different shapes and sizes. SQL will always be used specifically to retrieve and review data in a tabular form but will never be the proper tool to visualize data—this task is best left to Python and its many visualization libraries, which will be the focus of the next chapter.

4
Visualizing Data with Python

Regardless of the field of work you operate in, the career path you've chosen, or the specific project you are working on, the ability to effectively communicate information to others will always be useful. In fact, exactly one hundred years ago, in 1921, Frederick R. Barnard first said something which has become a phrase you have probably heard countless times: *A picture is worth a thousand words.*

With the many new technologies that have emerged in the realm of machine learning in recent years, the amount of data being structured, processed, and analyzed has grown exponentially. The ability to take data in its raw form and translate it to a meaningful and communicative diagram is one of the most sought-after skill sets in the industry today. Most decisions made in large companies and corporations are generally data-driven, and the best way to start a conversation about an area you care about is to create a meaningful visualization about it. Consider the following:

- The human brain is able to process visualizations 60,000 times faster than text.

- Nearly 90% of all information processed by the human brain is done visually.

- Visualizations are 30 times more likely to be read than even simple text.

Visualizations are not always about driving a conversation or convincing an opposing party to agree on something – they are often used as a means to investigate and explore data for the purposes of uncovering hidden insights. In almost every machine learning project you undertake, a significant amount of effort will be devoted to exploring data to uncover its hidden **features** through a process known as **Exploratory Data Analysis (EDA)**. EDA is normally done prior to any type of machine learning project in order to better understand the data, its features, and its limits. One of the best ways to explore data in this fashion is in a visual form, allowing you to uncover much more than the numerical values alone.

Over the course of the following chapter, we will look over some useful steps to follow to develop a robust visual for a given dataset. We will also explore some of the most common visualization libraries used in the **Python** community today. Finally, we will explore several datasets and learn how to develop some of the most common visualizations for them.

Within this chapter, we will cover the following main topics:

- Exploring the six steps of data visualization
- Commonly used visualization libraries
- Tutorial – visualizing data in Python

Technical requirements

In this chapter, we will apply our understanding of Python and **Structured Query Language** (**SQL**) to retrieve data and design meaningful visualizations through a number of popular Python libraries. These libraries can be installed using the `pip` installer demonstrated in *Chapter 2, Introducing Python and the Command Line*. Recall that the process of installing a library is done via the command line:

```
$ pip install library-name
```

So, now that we are set up, let's begin!

Exploring the six steps of data visualization

When it comes to effectively communicating key trends in your data, the method in which it is presented will always be important. When presenting any type of data to an audience, there are two main considerations: first, selecting the correct segment of data for the argument; second, selecting the most effective visualization for the argument. When working on a new visualization, there are six steps you can follow to help guide you:

1. **Acquire**: Obtain the data from its source.
2. **Understand**: Learn about the data and understand its categories and features.
3. **Filter**: Clean the data and remove missing values, NaN values, and corrupt entries.
4. **Mine**: Identify patterns or engineer new features.
5. **Condense**: Isolate the most useful features.
6. **Represent**: Select a representation for these features.

Let's look at each step in detail.

The first step is to *acquire* your data from its source. This source may be a simple CSV file, a relational database, or even a **NoSQL** database.

Second, it is important to *understand* the context of the data as well as its content. As a data scientist, your objective is to place yourself in the shoes of your stakeholders and understand their data as best you can. Often, a simple conversation with a **Subject Matter Expert** (**SME**) can save you hours by highlighting facts about the data that you otherwise would not have known.

Third, *filtering* your data will always be crucial. Most real-world applications of data science rarely involve ready-to-use datasets. Often, data in its raw form will be the main data source, and it is up to data scientists and developers to ensure that any missing values and corrupt entries are taken care of. Data scientists often refer to this step as **preprocessing**, and we will explore this in more detail in *Chapter 5, Understanding Machine Learning*.

With the data preprocessed, our next objective is to *mine* the data in an attempt to identify patterns or engineer new features. In simple datasets, values can often be quickly visualized as either increasing or decreasing, allowing us to easily understand the trend. In multidimensional datasets, these trends are often more difficult to uncover. For example, a time-series graph may show you an increasing *trend*, however, the first derivative of this graph may expose *trends* relating to *seasonality*.

Once a trend of interest is identified, the data representing that trend is often *isolated* from the rest of the data. And finally, this trend is *represented* using a visualization that complements it.

It is important to understand that these steps are by no means hard rules, but they should be thought of as useful guidelines to assist you in generating effective visualizations. Not every visualization will require every step! In fact, some visualizations may require other steps, perhaps sometimes in a different order. We will go through a number of these steps to generate some visualizations later in the *Tutorial – Visualizing data in Python* section within this chapter. When we do, try to recall these steps and see if you can identify them.

Before we begin generating some interesting visuals, let's talk about some of the libraries we will need.

Commonly used visualization libraries

There are countless **visualization libraries** available in Python, and more are being published every day. Visualization libraries can be divided into two main categories: **static visualization** libraries and **interactive visualization** libraries. Static visualizations are images consisting of plotted values that cannot be clicked by the user. On the other hand, interactive visualizations are not just images but representations that can be clicked on, reshaped, moved around, and scaled in a particular direction. Static visualizations are often destined for email communications, printed publications, or slide decks, as they are visualizations that you do not intend others to change. However, interactive visualizations are generally destined for dashboards and websites (such as **AWS** or **Heroku**) in anticipation of users interacting with them and exploring the data as permitted.

The following open source libraries are currently some of the most popular in the industry. Each of them has its own advantages and disadvantages, which are detailed in the following table:

Library Name	Type	Best for
matplotlib	Static	Quick & Easy Plots & Graphs
seaborn	Static	Aesthetically Pleasing Visuals
plotly	Interactive	Interactivity and Hover-over capabilities
Bokeh	Interactive	Real-time data visualizations

Figure 4.1 – A list of the most common visualization libraries in Python

Now that we know about visualization libraries, let's move on to the next section!

Tutorial – visualizing data in Python

Over the course of this tutorial, we will be retrieving a few different datasets from a range of sources and exploring them through various kinds of visualizations. To create these visuals, we will implement many of the visualization steps in conjunction with some of the open source visualization libraries. Let's get started!

Getting data

Recall that, in *Chapter 3, Getting Started with SQL and Relational Databases*, we used AWS to create and deploy a database to the cloud, allowing us to query data using **MySQL Workbench**. This same database can also be queried directly from Python using a library known as `sqlalchemy`:

1. Let's query that dataset directly from **Amazon Relational Database Service (RDS)**. To do so, we will need the `endpoint`, `username`, and `password` values generated in the previous chapter. Go ahead and list these as variables in Python:

```
ENDPOINT=" yourEndPointHere>"
PORT="3306"
USR="admin"
DBNAME="toxicity_db_tutorial"
PASSWORD = "<YourPasswordHere>"
```

2. With the variables populated with your respective parameters, we can now query this data using `sqlalchemy`. Since we are interested in the dataset as a whole, we can simply run a `SELECT * FROM dataset_toxicity_sd` command:

```
from sqlalchemy import create_engine
import pandas as pd
db_connection_str =
'mysql+pymysql://{USR}:{PASSWORD}@{ENDPOINT}:{PORT}/
{DBNAME}'.format(USR=USR, PASSWORD=PASSWORD,
ENDPOINT=ENDPOINT, PORT=PORT, DBNAME=DBNAME)
db_connection = create_engine(db_connection_str)
df = pd.read_sql('SELECT * FROM dataset_toxicity_sd',
con=db_connection)
```

Alternatively, you can simply import the same dataset as a CSV file using the `read_csv()` function:

```
df = pd.read_csv("../../datasets/dataset_toxicity_
sd.csv")
```

3. We can take a quick look at the dataset to understand its content using the `head()` function. Recall that we can choose to reduce the columns by specifying the names of the ones we are interested in by using double square brackets (`[ [] ]`):

```
df[["ID", "smiles", "toxic"]].head()
```

This gives us the following output:

	ID	smiles	toxic
0	25239916	c1c2c(c(c(c1[131I])[O-])[131I])Oc3c(cc(c(c3[13...	0
1	25239917	CCC[C@@H]1C[C@H]([NH+](C1)C)C(=O)N[C@@H]([C@@H...	0
2	25239918	CNC(=O)c1cc(ccn1)Oc2ccc(cc2)NC(=O)Nc3ccc(c(c3)...	0
3	25239919	CN(C)c1cccc2c1ccc(c2)S(=O)(=O)[O-]	0
4	25239920	CC(C)c1ccc2c(c1)c(=O)c3cc(c(nc3o2)N)C(=O)[O-]	0

Figure 4.2 – A DataFrame representation of selected columns from the toxicity dataset

If you recall, there are quite a few columns within this dataset, ranging from general data such as the primary key (ID) to the structure (smiles) and the toxicity (toxic). In addition, there are many features that describe and represent the dataset, ranging from the total polar surface area (TPSA) all the way to lipophilicity (LogP).

4. We can also get a sense of some of the general statistics behind this dataset – such as the maximum, minimum, and averages relating to each column – by using the describe() function in pandas:

```
df[["toxic", "TPSA", "MolWt", "LogP"]].describe()
```

This results in the following table:

	toxic	TPSA	MolWt	LogP
count	1460.000000	1460.000000	1460.000000	1460.000000
mean	0.064384	95.362767	382.356525	1.292078
std	0.245519	89.443235	228.985999	3.163150
min	0.000000	0.000000	27.026000	-19.396500
25%	0.000000	44.750000	253.275000	-0.156250
50%	0.000000	75.270000	334.350500	1.550400
75%	0.000000	112.625000	440.557250	3.143200
max	1.000000	833.780000	1882.332000	12.605800

Figure 4.3 – Some general statistics of selected columns from the toxicity dataset

Immediately, we notice that the **means** and **standard deviations** of each of the columns are drastically different from each other. We also notice that the minimum and maximum values are also quite different, in the sense that many of the columns are *integers* (whole numbers), whereas others are *floats* (decimals). We can also see that many of the minimums have values of zero, with two columns (FormalCharge and LogP) having negative values. So, this real-world dataset is quite diverse and spread out.

5. Before we can explore the dataset further, we will need to ensure that there are no missing values. To do this, we can run a quick check using the isna() function provided by the pandas library. We can chain this with the sum() function to get a sum of all of the missing values for each column:

```
df.isna().sum()
```

The result is shown in *Figure 4.4*:

```
ID                  0
smiles              0
toxic               0
FormalCharge        0
TPSA                0
MolWt               0
HeavyAtoms          0
NHOH                0
HAcceptors          0
HDonors             0
Heteroatoms         0
AromaticRings       0
SaturatedRings      0
AromaticOH          0
AromaticN           0
LogP                0
dtype: int64
```

Figure 4.4 – The list of missing values within the DataFrame

Thankfully, there are no missing values from this particular dataset, so we are free to move forward with creating a few plots and visuals.

Important note

Missing values: Please note that missing values can be addressed in a number of different ways. One option is to exclude a row with missing values from the dataset completely by using the dropna() function. Another option is to replace any missing value with a common value using the fillna() or replace() functions. Finally, you can also replace missing values with the mean of all the other values using the mean() function. The method you select will be highly dependent on the identity and meaning of the column.

Summarizing data with bar plots

Bar plots or **bar charts** are often used to describe *categorical data* in which the lengths or heights of the bars are proportional to the values of the categories they represent. Bar plots provide a visual estimate of the central tendency of a dataset with the uncertainty of the estimate represented by error bars.

So, let's create our first bar plot. We will be using the `seaborn` library for this particular task. There are a number of different ways to style your graphs. For the purposes of this tutorial, we will use the `darkgrid` style from `seaborn`.

Let's plot the `TPSA` feature relative to the `FormalCharge` feature to get a sense of the relationship between them:

```
import pandas as pd
import seaborn as sns
plt.figure(figsize=(10,5))
sns.barplot(x="FormalCharge", y="TPSA", data=df);
```

Our initial results are shown in *Figure 4.5*:

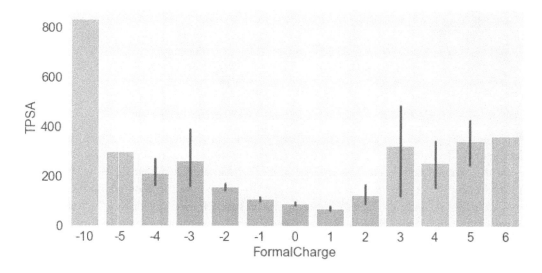

Figure 4.5 – A bar plot of the TPSA and FormalCharge features

Immediately, we can see an interesting relationship between the two, in the sense that the TPSA feature tends to increase when the absolute value of FormalCharge is further away from zero. If you are following along with the provided **Jupyter notebooks**, feel free to explore a few other relationships within this dataset. Now, let's try exploring the number of hydrogen donors (HDonors) instead of TPSA:

```
sns.barplot(x="FormalCharge", y="HDonors", data=df)
```

We can see the subsequent output in *Figure 4.6*:

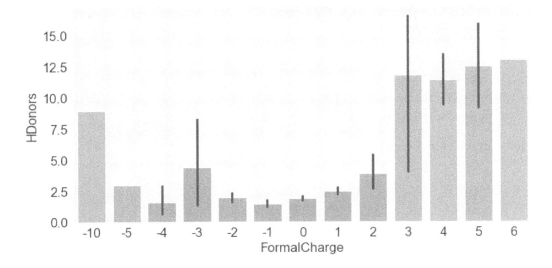

Figure 4.6 – A bar plot of the HDonors and FormalCharge features

Taking a look at the plot, we do not see as strong a relationship between the two variables. The highest and lowest formal charges do in fact show higher hydrogen donors. Let's compare this to HAcceptors – a similar feature in this dataset. We could either plot this feature individually, as we did with the hydrogen donors, or we could combine them both into one diagram. We can do this by *isolating* the features of interest (do you remember the name of this step?) and then *reshaping* the dataset. DataFrames within Python are often **reshaped** using four common functions:

Figure 4.7 – Four of the most common DataFrame reshaping functions

Each of these functions serves to reshape the data in a specific way. The pivot() function is often used to reshape a DataFrame organized by its index. The stack() function is often used with multi-index DataFrames – this allows you to *stack* your data, making the table *long and narrow* instead of *wide and short*. The melt() function is similar to the stack() function in the sense that it also *stacks* your data, but the difference between them is that stack() will insert the compressed columns into the inner index, whereas melt() will create a new column called Variable. Finally, unstack() is simply the opposite of stack(), in the sense that data is converted from *long* to *wide*.

For the purposes of comparing the hydrogen donors and acceptors, we will be using the melt() function, which you can see in *Figure 4.8*. Note that two new columns are created in the process: Variable and Value:

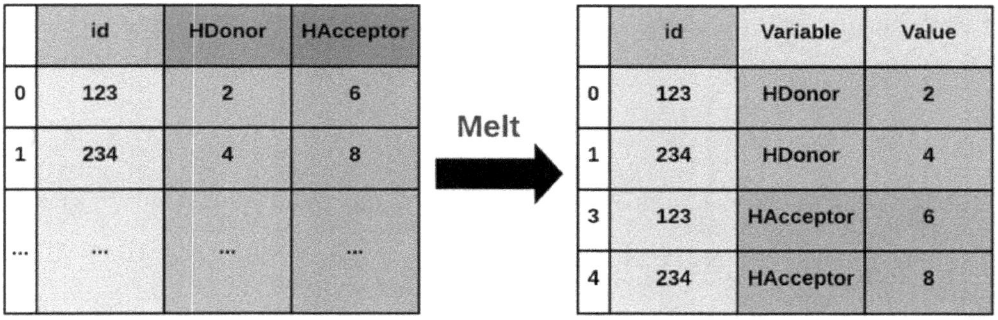

Figure 4.8 – A graphical representation of the melt() function

First, we create a variable called df_iso to represent the isolated DataFrame, and then we use the melt() function to *melt* its data and assign it to a new variable called df_melt. We can also print the shape of the data to prove to ourselves that the columns *stack* correctly if they exactly *double* in length. Recall that you can also check the data using the head() function:

```
df_iso = df[["FormalCharge", "HDonors", "HAcceptors"]]
print(df_iso.shape)
    (1460, 3)
df_melted = pd.melt(df_iso, id_vars=["FormalCharge"],
                value_vars=["HDonors", "HAcceptors"])
print(df_melted.shape)
    (2920, 3)
```

Finally, with the data ordered correctly, we can go ahead and plot this data, specifying the x-axis as `FormalCharge`, and the y-axis as `value`:

```
sns.barplot(data=df_melted, x='FormalCharge', y='value',
            hue='variable')
```

Upon executing this line of code, we will get the following figure:

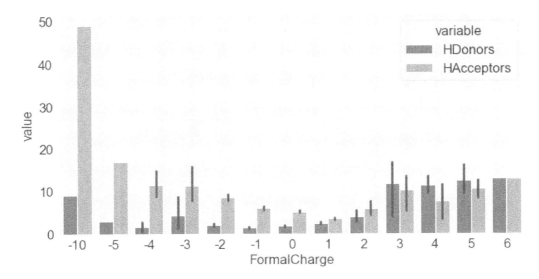

Figure 4.9 – A bar plot of two features relative to FormalCharge

As you begin to explore the many functions and classes within the `seaborn` library, referring to the documentation as you write your code can help you to debug errors and also uncover new functionality that you may not have known about. You can view the Seaborn documentation at `https://seaborn.pydata.org/api.html`.

Working with distributions and histograms

Histograms are plots that visually portray a summary or approximation of the distribution of a *numerical dataset*. In order to construct a histogram, **bins** must be established, representing a range of values by which the full range is divided. For example, take the molecular weights of the items in a dataset; we can plot a histogram of the weights in bins of 40:

```python
plt.figure(figsize=(10,5))
plt.title("Histogram of Molecular Weight (g/mol)", fontsize=20)
plt.xlabel("Molecular Weight (g/mol)", fontsize=15)
plt.ylabel("Frequency", fontsize=15)
df["MolWt"].hist(figsize=(10, 5),
                 bins=40,
                 xlabelsize=10,
                 ylabelsize=10,
                 color = "royalblue")
```

We can see the output of this code in *Figure 4.10*:

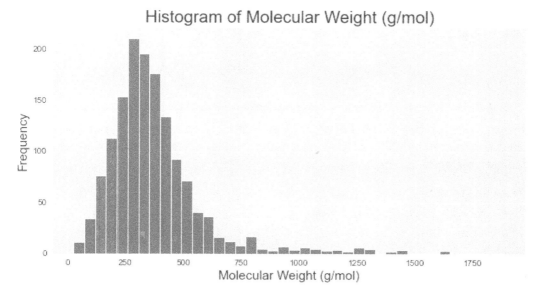

Figure 4.10 – A histogram of molecular weight with a bin size of 40

As you explore more visualization methods in Python, you will notice that most libraries offer a number of quick functions that have already been developed and optimized to perform a specific task. We could go through the same process of reshaping our data for each feature and iterate through them to plot a histogram for each of the features, or we could simply use the `hist()` function for them collectively:

```
dftmp = df[["MolWt", "NHOH", "HAcceptors", "Heteroatoms",
            "LogP", "TPSA"]]
dftmp.hist(figsize=(30, 10), bins=40, xlabelsize=10,
           ylabelsize=10, color = "royalblue")
```

The subsequent output can be seen in *Figure 4.11*:

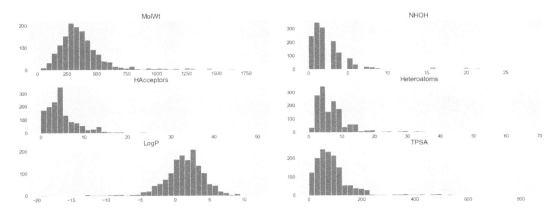

Figure 4.11 – A series of histograms for various features automated using the hist() function

Histograms can also be overlaid in order to showcase two features on the same plot. When doing this, we would need to give the plots a degree of transparency by using the `alpha` parameter:

```
dftmp = df[["MolWt","TPSA"]]
x1 = dftmp.MolWt.values
x2 = dftmp.TPSA.values
kwargs = dict(histtype='stepfilled', alpha=0.3,
              density=True, bins=100, ec="k")
plt.figure(figsize=(10,5))
plt.title("Histogram of Molecular Weight (g/mol)",
          fontsize=20)
```

```
plt.xlabel("Molecular Weight (g/mol)", fontsize=15)
plt.ylabel("Frequency", fontsize=15)
plt.xlim([-100, 1000])
plt.ylim([0, 0.01])
plt.hist(x1, **kwargs)
plt.hist(x2, **kwargs)
plt.legend(dftmp.columns)
plt.show()
```

We can see the output of the preceding command in *Figure 4.12*:

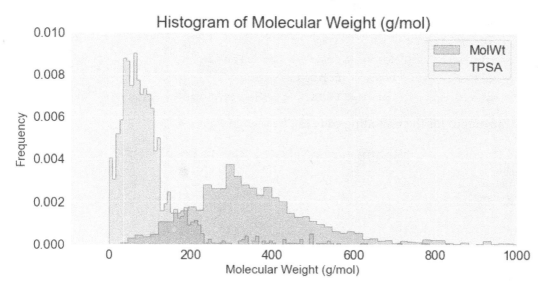

Figure 4.12 – An overlay of two histograms where their opacity was reduced

Histograms are wonderful ways to summarize and visualize data in large quantities, especially when the functionality is as easy as using the `hist()` function. You will find that most libraries – such as `pandas` and `numpy` – have numerous functions with similar functionality.

Visualizing features with scatter plots

Scatter plots are representations based on *Cartesian coordinates* that allow for visualizations to be created in both two- and three-dimensional spaces. Scatter plots consist of an x-axis and a y-axis and are normally accompanied by an additional feature that allows for separation within the data. Scatter plots are best used when accompanied by a third feature that can be represented either by color or shape, depending on the data type available. Let's look at a simple example:

1. We'll take a look at an example of a simple scatter plot showing TPSA relative to the HeavyAtoms feature:

```
plt.figure(figsize=(10,5))
plt.title("Scatterplot of Heavy Atoms and TPSA",
fontsize=20)
plt.ylabel("Heavy Atoms", fontsize=15)
plt.xlabel("TPSA", fontsize=15)
sns.scatterplot(x="TPSA", y="HeavyAtoms", data=df)
```

The output for the preceding code can be seen in *Figure 4.13*:

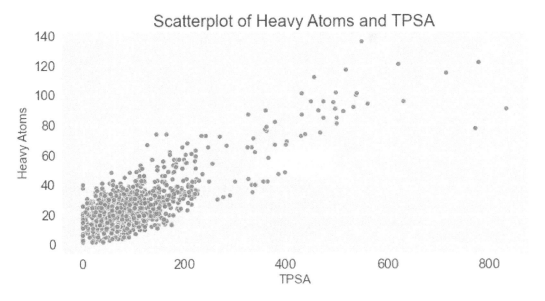

Figure 4.13 – A scatter plot of the TPSA and HeavyAtoms features

Immediately, we notice that there is some dependency between the two features, as shown by the slight positive correlation.

2. We can take a look at a third feature, such as `MolWt`, by changing the color and size using the `hue` and `size` arguments, respectively. This gives us the ability to plot three or four features on the same graph, giving us an excellent interpretation of the dataset. We can see some trending among `TPSA` relative to `HeavyAtoms`, and increasing `MolWt`:

```
plt.figure(figsize=(10,5))
plt.title("Scatterplot of Heavy Atoms and TPSA",
fontsize=20)
plt.ylabel("Heavy Atoms", fontsize=15)
plt.xlabel("Molecular Weight (g/mol)", fontsize=15)
sns.scatterplot(x="TPSA",y="HeavyAtoms",
size="MolWt", hue="MolWt", data=df)
```

The output of the preceding code can be seen in *Figure 4.14*:

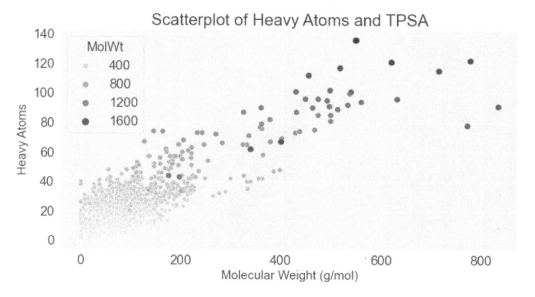

Figure 4.14 – A scatter plot of two features, with a third represented by size and color

3. As an alternative to 2D scatter plots, we can use 3D scatter plots to introduce another feature in the form of a new dimension. We can take advantage of the `Plotly` library to implement some 3D functionality. To do this, we can define a `fig` object using the `scatter_3d` function, and subsequently, we define the source of our data and the axes of interest:

```
import plotly.express as px
fig = px.scatter_3d(df, x='TPSA', y='LogP',
z='HeavyAtoms',
                         color='toxic', opacity=0.7)
fig.update_traces(marker=dict(size=4))
fig.show()
```

The output of this code will result in *Figure 4.15*:

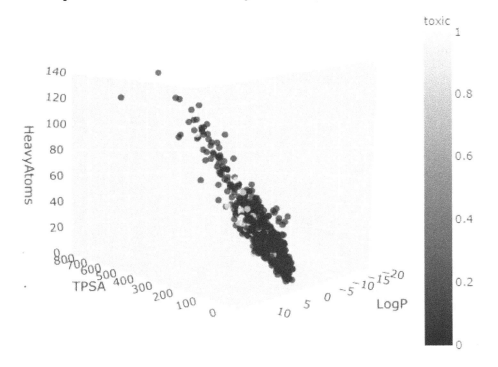

Figure 4.15 – A 3D scatter plot of three features, colored by toxicity

4. Instead of adding more features, we can add some more elements to the scatter plot to help interpret the two features on the x and y coordinates. We noticed earlier that there was a slight correlation within the dataset that seems ripe for exploration. It would be interesting to see if this correlation holds true for both toxic and non-toxic compounds. We can get a sense of the correlation using the `lmplot()` function, which allows us to graphically represent the correlation as a *linear regression* within the scatter plot:

```
sns.lmplot(x="HAcceptors", y="TPSA", hue="toxic",
           data=df, markers=["o", "x"], height = 5,
           aspect = 1.7, palette="muted");
plt.xlim([0, 16])
plt.ylim([0, 400])
```

The subsequent output can be seen in *Figure 4.16*:

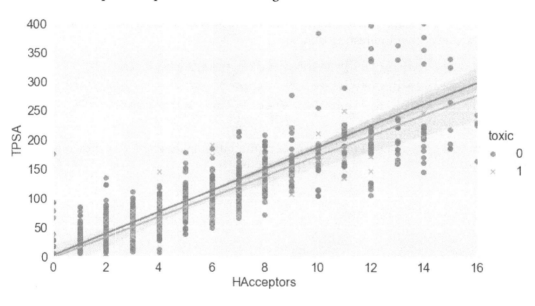

Figure 4.16 – A scatter plot of two features and their associated correlations

Scatter plots are great ways to portray data relationships and begin to understand any dependencies or correlations they may have. Plotting regressions or lines of best fit can give you some insight into any possible relationships. We will explore this in greater detail in the following section.

Identifying correlations with heat maps

Now that we have established a correlation between two molecular features within our dataset, let's investigate to see if there are any others. We can easily go through each set of features, plot them, and look at their respective regressions to determine whether or not a correlation may exist. In Python, automating whenever possible is advised, and luckily for us, this task has already been automated! So, let's take a look:

1. Using the `pairplot()` function will take your dataset as input and return a figure of all the scatter plots for all of the features within your dataset. To fit the figure within the confines of this page, only the most interesting features were selected. However, I challenge you to run the code in the provided Jupyter notebook to see if there are any other interesting trends:

    ```
    featOfInterest = ["TPSA", "MolWt", "HAcceptors",
            "HDonors", "toxic", "LogP"]
    sns.pairplot(df[featOfInterest], hue = "toxic",
    markers="o")
    ```

    The results are presented in the form of numerous smaller graphs, as shown in *Figure 4.17*:

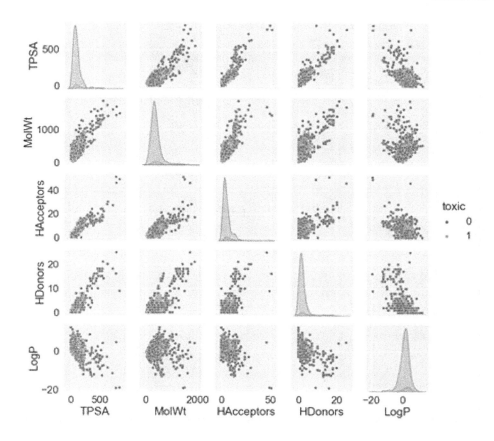

Figure 4.17 – A pairplot() graphic of the toxicity dataset for selected features

2. Alternatively, we can capture the *Pearson correlation* for each of the feature pairs using the `corr()` function in conjunction with the DataFrame itself:

```
df[["TPSA", "MolWt", "HeavyAtoms", "NHOH", "HAcceptors",
        "HDonors", "AromaticRings", "LogP",
  "AromaticN"]].corr()
```

We can review these correlations as a DataFrame in *Figure 4.18*:

	TPSA	MolWt	HeavyAtoms	NHOH	HAcceptors	HDonors	AromaticRings	LogP	AromaticN
TPSA	1.000000	0.785735	0.781096	0.823539	0.897188	0.849590	0.219956	-0.540883	0.195574
MolWt	0.785735	1.000000	0.961451	0.609623	0.760046	0.686076	0.397915	-0.038773	0.122023
HeavyAtoms	0.781096	0.961451	1.000000	0.614842	0.755139	0.688060	0.440247	-0.012894	0.152883
NHOH	0.823539	0.609623	0.614842	1.000000	0.570108	0.963666	0.160302	-0.562992	0.072826
HAcceptors	0.897188	0.760046	0.755139	0.570108	1.000000	0.633044	0.237554	-0.392125	0.282243
HDonors	0.849590	0.686076	0.688060	0.963666	0.633044	1.000000	0.210288	-0.505401	0.085185
AromaticRings	0.219956	0.397915	0.440247	0.160302	0.237554	0.210288	1.000000	0.242826	0.522242
LogP	-0.540883	-0.038773	-0.012894	-0.562992	-0.392125	-0.505401	0.242826	1.000000	-0.035307
AromaticN	0.195574	0.122023	0.152883	0.072826	0.282243	0.085185	0.522242	-0.035307	1.000000

Figure 4.18 – A DataFrame showing the correlations between selected features

3. For a more visually appealing result, we can *wrap* our data within a heatmap() function and apply a color map to show dark colors for strong correlations and light colors for weaker ones:

```
sns.heatmap(df[["TPSA", "MolWt", "HeavyAtoms", "NHOH",
            "HAcceptors", "HDonors", "AromaticRings",
            "LogP", "AromaticN"]].corr(),
            annot = True,  cmap="YlGnBu")
```

Some of the code we have written so far has become a little complicated as we begin to *chain* multiple functions together. To provide some clarity of the syntax and structure, let's take a closer look at the following function. We begin by calling the main heatmap class within the seaborn library (recall that we give this the alias sns). We then add our dataset, containing the sliced set of the features of interest. We then apply the correlation function to get the respective correlations, and finally add some additional arguments to style and color the plot:

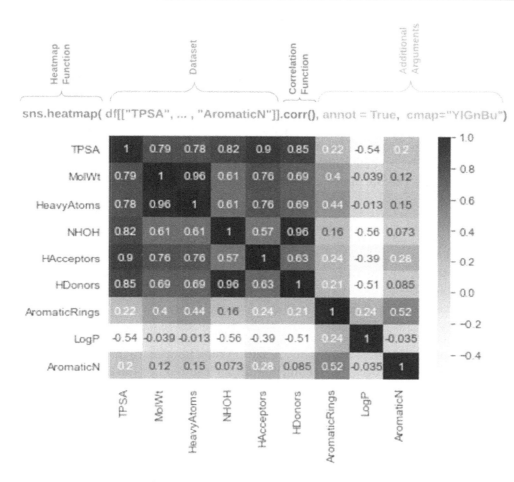

Figure 4.19 – A heat map showing the correlation between selected features

Identifying correlations within datasets will always be useful, regardless of whether you are analyzing data or preparing a predictive model. You will find that `corr()` and many of its derivatives are commonly used in the machine learning space.

Displaying sequential and time-series plots

The datasets and features we have explored so far have all been provided in a *structured* and *tabular* form, existing as rows and columns within DataFrames. These rows are fully independent of each other. This is not always the case in all datasets, and *dependence* (especially *time-based dependence*) is sometimes a factor we need to consider. For example, take a **Fast All (FASTA)** sequence – that is, a text-based format often used in the realm of bioinformatics for representing nucleotide or amino acid sequences via letter codes. In molecular biology and genetics, a parameter known as **Guanine-Cytosine (GC) content** is a metric used to determine the percent of nitrogenous bases within DNA or RNA molecules. Let's explore plotting this sequential data using a FASTA file for COVID-19 data:

1. We will begin the process by importing the dataset using the wget library:

```
import wget
url_covid = "https://ftp.expasy.org/databases/uniprot/
pre_release/covid-19.fasta"
filename = wget.download(url_covid, out="../../datasets")
```

2. Next, we can calculate the GC content using the Biopython (also called Bio) library – one of the most commonly utilized Python libraries in the computational molecular biology space. The documentation and tutorials for the Biopython library can be found at http://biopython.org/DIST/docs/tutorial/Tutorial.html.

3. We will then parse the file using the SeqIO and GC classes and write the results to the gc_values_covid variable:

```
from Bio import SeqIO
from Bio.SeqUtils import GC
gc_values_covid = sorted(GC(rec.seq) for rec in
    SeqIO.parse("../../datasets/covid-19.fasta",
"fasta"))
```

Please note that the path to the file in the preceding code may change depending on which directory the file was saved in.

4. Finally, we can go ahead and plot the results using either pylab or matplotlib:

```
import pylab
plt.figure(figsize=(10,5))
plt.title("COVID-19 FASTA Sequence GC%", fontsize=20)
plt.ylabel("GC Content %", fontsize=15)
```

```
plt.xlabel("Genes", fontsize=15)
pylab.plot(gc_values_covid)
pylab.show()
```

The subsequent output can be seen in *Figure 4.20*:

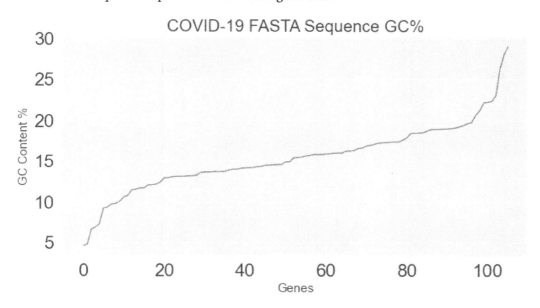

Figure 4.20 – A plot showing the GC content of the COVID-19 sequence

While there are many non-time-based sequential datasets such as `text`, `images`, and `audio`, there are also time-based datasets such as `stock prices` and `manufacturing processes`. Within the laboratory space, there are many pieces of equipment that also utilize time series-based approaches, such as those relating to chromatography. For example, take **Size-Exclusion Chromatography** (**SEC**), in which molecules are separated by their sizes. This property is known as *molecular weight*. Most predictive maintenance models tend to monitor temperature and pressure to detect anomalies and alert users. Let's go ahead and pull in the following `time-series` dataset and overlay `Temperature` and `Pressure` together over time:

```
dfts = pd.read_csv("../../datasets/dataset_pressure_ts.csv")
plt.title("Timeseries of an LCMS Chromatogram (Pressure &
    Temperature)", fontsize=20)
plt.ylabel("Pressure (Bar)", fontsize=15)
plt.xlabel("Run Time (min)", fontsize=15)
ax1 = sns.lineplot(x="Run Time", y="Pressure",
```

```
                    data=dfts, color = "royalblue",
                    label = "Pressure (Bar)");
ax2 = sns.lineplot(x="Run Time", y="Temperature",
                    data=dfts, color = "orange",
                    label = "Pressure (Bar)");
```

The output of this code can be seen in *Figure 4.21*:

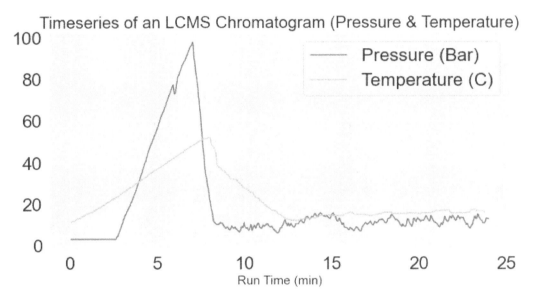

Figure 4.21 – A time-series plot showing the temperature and pressure of a failed LCMS run

We notice that within the first 5 minutes of this graph, the temperature and pressure parameters are increasing quite quickly. A dip of some sort occurs within the 6.5-minute range, and the system keeps increasing for a moment, then both parameters begin to plummet downward and level out at their respective ranges. This is an example of an instrument failure, and it is a situation that a finely tuned machine learning model would be able to detect relative to its successful counterpart. We will explore the development of this anomaly detection model in greater detail in *Chapter 7, Supervised Machine Learning*.

Emphasizing flows with Sankey diagrams

A popular form of visualization in data science is the **Sankey diagram** – made famous by Minard's classic depiction of Napoleon's army during the invasion of Russia. The main purpose of a Sankey diagram is to visualize a magnitude in terms of its proportional width on a flow diagram:

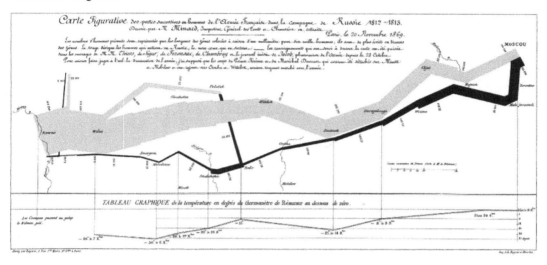

Figure 4.22 – A Sankey diagram by Charles Joseph Minard depicting Napoleon's march to Russia

Sankey diagrams are often used to depict many applications across various sectors. Biotechnology and health sector applications of Sankey diagrams include the following:

- Depictions of drug candidates during clinical trials
- Process flow diagrams for synthetic molecules
- Process flow diagrams for microbial fermentation
- Project flow diagrams and success rates
- Financial diagrams depicting costs within an organization

Let's visualize a simple example of a company's drug candidate pipeline. We'll take the total number of candidates, their classification by phase, and finally, their designation by modality as small or large molecules. We can take advantage of the `Plotly` library to assist us with this:

```
import plotly.graph_objects as go
fig = go.Figure(data=[go.Sankey(node = dict(pad = 50,
        thickness = 10,
                line = dict(color = "black", width = 0.5),
                label = ["Drug Candidates", "Phase 1", "Phase
2",
                "Phase 3", "Small Molecules", "Large
Molecules"],
                color = "blue"),
                link = dict(
                source = [0,   0, 0, 1,   2, 3, 1, 2, 3],
                target = [1,   2, 3, 4,   4, 4, 5, 5, 5],
                value = [15, 4, 2, 13, 3, 1, 2, 1, 1]
    ))])
```

This segment of code is quite long and complex – let's try to break this down. The `figure` object consists of several arguments we need to take into account. The first is pad, which describes the spacing between the *nodes* of the visualization. The second describes the `thickness` value of the node's bars. The third sets the `color` and `width` values of the lines. The fourth contains the `label` names of the nodes. And finally, we arrive at the data, which has been structured in a slightly different way to how we are accustomed. In this case, the dataset is divided into a `source` array (or origin), the `target` array, and the `value` array associated with it. Starting on the left-hand side, we see that the first value of `source` is node 0, which goes to the `target` of node 1, with a `value` of 15. Reading the process in this fashion makes the flow of the data a little clearer to the user or developer. Finally, we can go ahead and plot the image using `show()`:

```
fig.update_layout(title_text="Drug Candidates within a Company
Pipeline", font_size=10)
fig.show()
```

The following diagram displays the output of the preceding code:

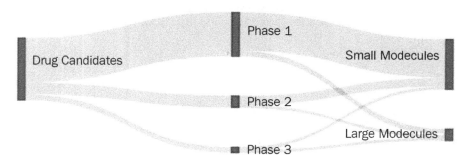

Figure 4.23 – A Sankey diagram representing a company's pipeline

Sankey diagrams are a great way to show the flow or transfer of information over time or by category. In the preceding example, we looked at its application in terms of small and large molecules within a pipeline. Let's now take a look at how we can visualize these molecules.

Visualizing small molecules

When it comes to small molecules, there are a number of ways we can visualize them using various software platforms and online services. Luckily, there exists an excellent library commonly utilized for **cheminformatics** applications known as the **Research and Development Kit** (**RDKit**) that allows for the depiction of small molecules using the **SMILES** format. The rdkit library can be installed using pip:

```
import pandas as pd
import rdkit
from rdkit import Chem
```

We can parse the DataFrame we imported earlier in this tutorial and extract a sample `smiles` string via indexing. We can then create a molecule object using the `MolFromSmiles()` function within the `Chem` class of `rdkit` using the `smiles` string as the single argument:

```
df = pd.read_csv("../../datasets/dataset_toxicity_sd.csv")
m = Chem.MolFromSmiles(df["smiles"][5])
m
```

The output of this variable can be seen in *Figure 4.24*:

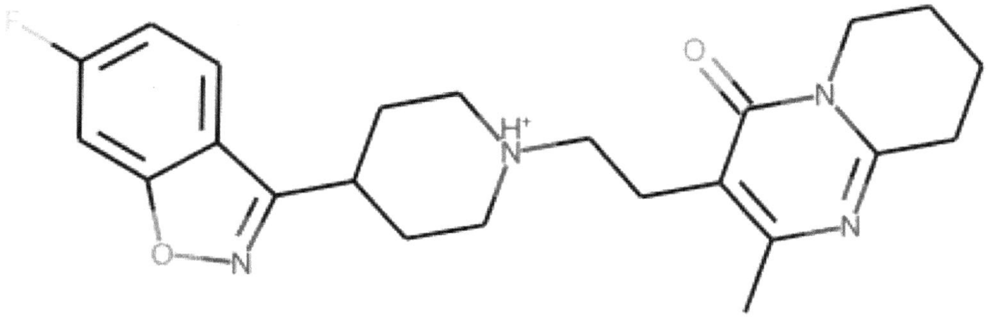

Figure 4.24 – A representation of a small molecule

We can check the structure of another molecule by looking at a different index value:

```
m = Chem.MolFromSmiles(df["smiles"][20])
m
```

This time, our output is as follows:

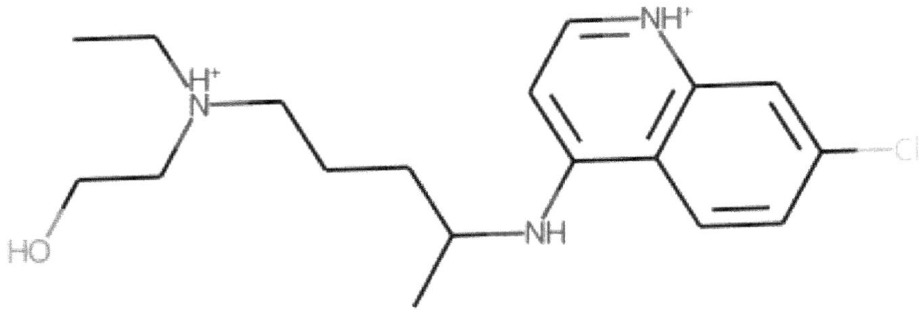

Figure 4.25 – A representation of a small molecule

In addition to rendering print-ready depictions of small molecules, the `rdkit` library also supports a wide variety of functions related to the analysis, prediction, and calculation of small molecule properties. In addition, the library also supports the use of charge calculations, as well as similarity maps:

```
from rdkit.Chem import AllChem
from rdkit.Chem.Draw import SimilarityMaps
AllChem.ComputeGasteigerCharges(m)
contribs = [m.GetAtomWithIdx(i).GetDoubleProp('_
GasteigerCharge') for i in range(m.GetNumAtoms())]
fig = SimilarityMaps.GetSimilarityMapFromWeights(m,
            contribs, contourLines=10, )
```

The output of the preceding code can be seen in *Figure 4.26*:

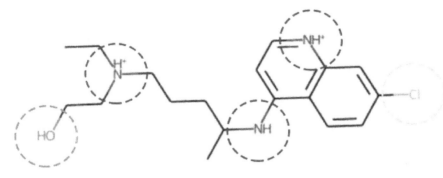

Figure 4.26 – A representation of a small molecule's charge

Now that we have gained an idea of how we can use RDKit to represent small molecules, let's look at an application of this for large molecules instead.

Visualizing large molecules

There are a number of Python libraries designed for the visualization, simulation, and analysis of large molecules for the purposes of research and development. Currently, one of the most common libraries is `py3Dmol`. Exclusively used for the purposes of 3D visualization within a Jupyter Notebook setting, this library allows for the creation of publication-ready visuals of 3D proteins. The library can be easily downloaded using the `pip` framework.

At the time of writing, the world is still in the midst of dealing with the COVID-19 virus that originated in Wuhan, China and spread throughout the world. On July 8, 2020, a 1.7 Å resolution structure of the *SARS-CoV-2 3CL* protease was released in the **RCSB Protein Data Bank (RCSB PDB)** at `pdb = 6XMK`. Let's go ahead and use this protein as an example in the following visualizations:

1. We can begin the development of this visual using the `py3dmol` library and querying the protein structure directly within the following function:

    ```
    import py3Dmol
    largeMol = py3Dmol.view(query='pdb:6xmk',
                            width=600,
                            height=600)
    ```

2. With the library imported, a new variable object called `lm` can be specified using the `view` class in `py3Dmol`. This function takes three main arguments. The first is the identity of the protein of interest, namely `6xmk`. The second and third arguments are the width and height of the display window, respectively. For more information about PDB files, visit the **RCSB PDB** at www.rcsb.org. Let's start by viewing this protein as a basic molecular stick structure by passing a `stick` argument:

    ```
    largeMol.setStyle({'stick':{'color':'spectrum'}})
    largeMol
    ```

Upon executing this line of code, we get the following image of the molecule:

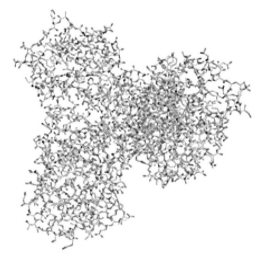

Figure 4.27 – A representation of a large molecule or protein in ball-stick form

3. Notice that we added a `stick` argument that displayed the last structure. We can change this argument to `cartoon` to see a cartoon representation of this protein based on its *secondary structure*:

```
largeMol.setStyle({'cartoon':{'color':'spectrum'}})
largeMol
```

When executing this line of code, we get the following image of the molecule:

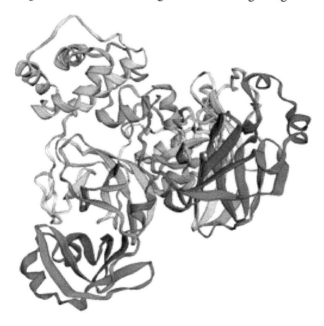

Figure 4.28 – A representation of a large molecule or protein's secondary structure

4. There are a number of other changes and arguments that can be added to custom fit this visualization to a user's particular aims. One of these changes is the addition of a **Van der Waals surface**, which allows for the illustration of the area through which a molecular interaction might occur. We will add this surface to only one of the two chains on this protein:

```
lm = py3Dmol.view(query='pdb:6xmk')
chA = {'chain':'A'}
chB = {'chain':'B'}
lm.setStyle(chA,{'cartoon': {'color':'spectrum'}})
lm.addSurface(py3Dmol.VDW, {'opacity':0.7,
'color':'white'}, chA)
```

```
lm.setStyle(chB,{'cartoon': {'color':'spectrum'}})
lm.show()
```

We can see the output of this code in *Figure 4.29*:

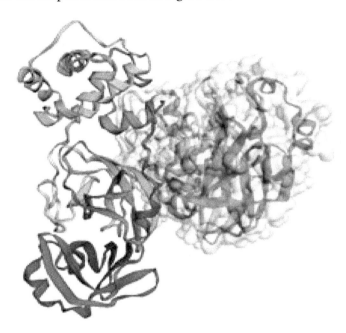

Figure 4.29 – A representation of a large molecule or protein's secondary structure with a Van der Waals surface on one of the chains

The study of large molecules, or **biologics**, have shown tremendous growth in the biotechnology sector in recent years. Within this chapter, we briefly introduced one of the many methods used to visualize these complex molecules – an important first step for any bioinformatics project.

Summary

Visualizations can be useful, powerful, and convincing tools to help illustrate points and drive conversations in specific directions. To create a proper visualization, there are certain steps and techniques that need to be taken to ensure your diagram is correct and effective.

Within this chapter, we explored the six main steps to follow when creating a proper visualization. We also explored many different methods and libraries within the scope of Python to help you create and style visuals for your specific aims. We explored some of the more basic visuals, such as bar plots, histograms, and scatter plots to analyze a few features at a time. We also explored more complex visualizations such as pair plots, heat maps, Sankey diagrams, and molecular representations, with which we can explore many more features.

We also touched on the concept of *correlation* and how certain features can have relationships with others – a concept we will cover in greater detail as we turn our attention to **machine learning** in the next chapter.

Section 2: Developing and Training Models

In this section, you will be introduced to the process of parsing data and training models. We begin with an introduction to the two forms of machine learning and reinforce the definitions of a number of modules/tutorials using real-world examples in the biotechnology industry.

This section comprises the following chapters:

- *Chapter 5, Understanding Machine Learning*
- *Chapter 6, Unsupervised Machine Learning*
- *Chapter 7, Supervised Machine Learning*
- *Chapter 8, Understanding Deep Learning*
- *Chapter 9, Natural Language Processing*
- *Chapter 10, Exploring Time Series Analysis*

5
Understanding Machine Learning

Over the last few years, you have likely heard some of the many popular buzz words such as **Artificial Intelligence (AI)**, **Machine Learning (ML)**, and **Deep Learning (DL)** that have rippled through most major industries. Although many of these phrases tend to be used interchangeably in company-wide all-staff and leadership meetings, each of these phrases does in fact refer to a distinct concept. So, let's take a look closer look at what these phrases actually refer to.

AI generally refers to the overarching domain of human-like intelligence demonstrated by software and machines. We can think of AI as the space that encompasses many of the topics we will discuss within the scope of this book.

Within the AI domain, there exists a sub-domain that we refer to as **machine learning**. ML can be defined as *the study of algorithms in conjunction with data to develop predictive models*.

Within the ML domain, there exists yet another sub-domain we refer to as **deep learning**. We will define **DL** as *the application of ML specifically through the use of artificial neural networks*.

Now that we have gained a better sense of the differences between these terms, let's define the concept of ML in a little more detail. There are several different definitions of ML that you will encounter depending on who you ask. Physicists tend to link the definition to applications in *performance optimization*, whereas mathematicians have a tendency to link the definition to *statistical probabilities* and, finally, computer scientists tend to link the definition to *algorithms* and *code*. To a certain extent, all three are technically correct. For the purposes of this book, we will define ML as a field of research concerning the development of mathematically optimized models using computer code, which *learn* or *generalize* from historical data to unlock useful insights and make predictions.

Although this definition may seem straightforward, most experienced interview candidates still tend to struggle when defining this concept. Make note of the exact phrasing we used here, as it may prove to be useful in a future setting.

Over the course of this chapter, we will visit various aspects of ML, and we will review some of the most common steps a developer must take when it comes to developing a **predictive model**.

In this chapter, we will review the following main topics:

- Understanding ML
- Overfitting and underfitting
- Developing an ML model

With all this in mind, let's get started!

Technical requirements

In this chapter, we will apply our understanding of **Python** to demonstrate some of the concepts behind ML. We will take a close look at some of the main steps in the ML model development process in which we will utilize some familiar libraries, such as `pandas` and `numpy`. In addition, we will also use some ML libraries such as `sklearn` and `tensorflow`. Recall that the process of installing a new library can be done via the command line:

```
$ pip install library-name
```

Let's begin!

Understanding ML

In the introduction, we broadly defined the concept of ML as it pertains to this book. With that definition in mind, let's now take a look at some examples to elaborate on our definition. In its broadest sense, ML can be divided into four areas: **classification**, **regression**, **clustering**, and **dimensionality reduction**. These four categories are often referred to as the field of **data science**. Data science is a very broad term used to refer to various applications relating to data, as well as the field of AI and its subsets. We can visualize the relationships between these fields in *Figure 5.1*:

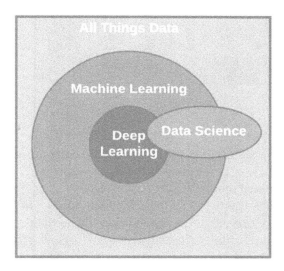

Figure 5.1 – The domain of AI as it relates to other fields

With these concepts in mind, let's discuss these four ML methods in more detail.

Classification is a method of *pattern detection* in which our objective is to predict a *label* (or *category*) from a finite set of possible options. For example, we can train a model to predict a protein's structure (for example, alpha helix or beta sheet), which can be referred to as a **binary classifier** as there are only *two* possible outcomes or categories. We can break down classification models even further; however, we will visit this in greater detail in *Chapter 7, Supervised Machine Learning*. For now, note that *classification* is a method to predict a label (or category). We begin with a dataset in which our input values (referred to as X) and their subsequent output values (generally referred to as ŷ) are used to train a classifier. This classifier can then be used to make predictions on new and unseen data. We can represent this visually in *Figure 5.2*:

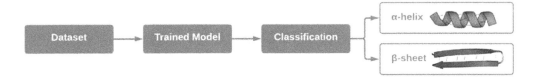

Figure 5.2 – An example of a classification model

Clustering is similar to classification in the sense that the outcome of the model is a label (or category), but the difference here is that a clustering model is not trained on a list of predefined classes but is based on the similarities between objects. The clustering model then groups the data points together in *clusters*. The total number of clusters formed is not always known ahead of time, and this depends heavily on the parameters the model is trained on. In the following example, three clusters were formed using the original dataset:

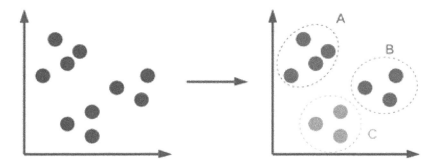

Figure 5.3 – An example of a clustering model

On the other hand, when it comes to **regression**, we are trying to predict a specific value, such as an **isoelectric point (pI)** in which the possible values are **continuous** (pI = 5.59, 6.23, 7.12, and so on). Unlike classification or clustering, there are no labels or categories involved here, only a numerical value. We begin with a dataset in which our input values (X) and their subsequent output values (ŷ) are used to train a *regressor*. This regressor can then be used to make predictions on new and unseen data:

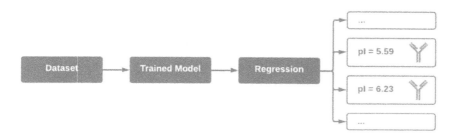

Figure 5.4 – An example of a regression model

Finally, when it comes to **dimensionality reduction**, ML can be applied not for the purposes of predicting a value, but in the sense of transforming data from a *high-dimensional* representation to a *low-dimensional* representation. Take, for example, the vast toxicity dataset we worked with in the previous chapters. We could apply a method such as **Principal Component Analysis (PCA)** to reduce the 10+ columns of features down to only two or three columns by *combining the importance* of these features together. We will examine this in greater detail in *Chapter 7, Understanding Supervised Machine Learning*. We can see a visual representation of this in *Figure 5.5*:

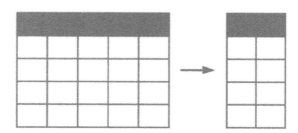

Figure 5.5 – An example of a dimensionality reduction model

The ML field is vast, complex, and extends well beyond the four basic examples we just touched on. However, the most common applications of ML models tend to focus on *predicting a category, predicting a value,* or *uncovering hidden insights* within data.

As scientists, we always want to organize our thoughts as best we can, and as it happens, the concepts we just discussed can be placed into two main categories: **Supervised Machine Learning** (**SML**) and **Unsupervised Machine Learning** (**UML**). SML encompasses all applications and models in which the datasets used to train the models contain both *features* and *ground truth* output values. In other words, we know both the input (X) and the output (ŷ). We call this a *supervised* method because the model was taught (*supervised*) which output label corresponds to which input value. On the other hand, UML encompasses ML models in which only the input (X) is known. Looking back to the four methods we discussed, we can divide them across both learning methods in the sense that **classification** and **regression** fall under SML, whereas **clustering** and **dimensionality reduction** fall under UML:

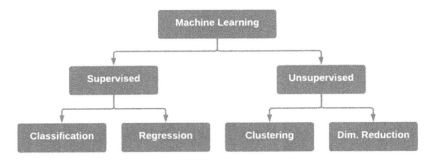

Figure 5.6 – A representation of supervised and unsupervised machine learning

Over the course of the following chapters, we will explore many popular ML models and algorithms that fall under these four general categories. As you follow along, I encourage you to develop a mind map of your own, and further branch out each of the four categories to all the different models you will learn. For example, we will explore a *Naïve Bayes* model in the *Saving a model for deployment* section of this chapter, which could be added to the **classification** branch of *Figure 5.6*. Perhaps you could branch out each of the models with some notes regarding the model itself. A map or visual aid may prove to be useful when preparing for a technical interview.

Throughout each of the models we develop, we will follow a particular set of steps to acquire our data, preprocess it, build a model, evaluate its performance, and finally, if the model is sufficient, deploy it to our end users or data engineers. Before we begin developing our models, let's discuss the common dangers known as *overfitting* and *underfitting*.

Overfitting and underfitting

Within the context of SML, we will prepare our models by *fitting* them with historical data. The process of fitting a model generally outputs a measure of how well the model generalizes to data that is similar to the data on which the model was trained. Using this output, usually in the form of **precision**, **accuracy**, and **recall**, we can determine whether the method we implemented or the parameters we changed had a positive impact on our model. If we revisit the definition of ML models that from earlier in this chapter, we specifically refer to them as models that *learn* or *generalize* from historical data. Models that are able to learn from historical data are referred to as *well-fitted* models, in the sense that they are able to perform accurately on new and unseen data.

There are instances in which models are underfitted. *Underfitted* models generally perform poorly on datasets, which means they have not learned to generalize well. These cases are generally the result of an inappropriate model being selected for a given dataset or the inadequate setting of a parameter/hyperparameter for that model.

> **Important note**
> **Parameters and hyperparameters**: Note that while parameters and hyperparameters are terms that are often used interchangeably, there is a difference between the two. *Hyperparameters* are parameters that are not learned by a model's estimator and must be manually tuned.

There are also instances in which models are overfitted. *Overfitted* models are models that *know* the dataset a little too well, and this means that they are no longer *learning* but *memorizing*. Overfitting generally occurs when a model begins to learn from the noise within a dataset and is no longer able to generalize well on new data. The differences between well-fitted, overfitted, and underfitted models can be seen in *Figure 5.7*:

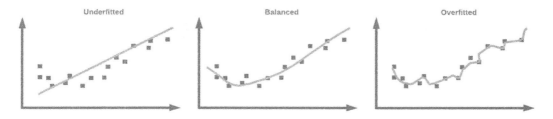

Figure 5.7 – A representation of overfitting and underfitting data

The objective of every data scientist is to develop a balanced model with optimal performance when it comes to your metrics of interest. One of the best ways to ensure that you are developing a balanced model that is not underfitting or overfitting is by splitting your dataset ahead of time and ensuring that the model is only ever trained on a subset of the data. We can split datasets into two categories: *training data* and *testing data* (also often referred to as *validation data*). We can use the training dataset to train the model, and we can use the testing dataset to test (or validate) the model. One of the most common classes to use for this purpose is the `train_test_split()` class from `sklearn`. If you think of your dataset with X being your input variables and $\hat{y}$ being your output, you can split the dataset using the following code snippet. First, we import the data. Then, we isolate the features we are interested in and output their respective variables. Then, we implement the `train_test_split()` function to split the data accordingly:

```
import pandas as pd
from sklearn.model_selection import train_test_split
df = pd.read_csv("../../datasets/dataset_wisc_sd.csv")
X = df.drop(columns = ["id", "diagnosis"])
y = df.diagnosis.values
X_train, X_test, y_train, y_test = train_test_split(X, y)
```

We can visualize the split dataset in *Figure 5.8*:

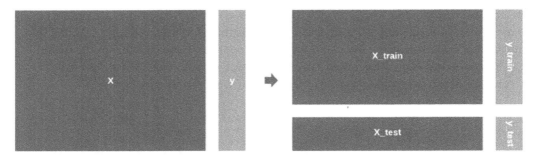

Figure 5.8 – A visual representation of data that has been split for training and testing

With the data split in this fashion, we can now use `X_train` and `y_train` for the purposes of training our model, and `X_test` and `y_test` for the purposes of testing (or validating) our model. The default splitting ratio is 75% training data to 25% testing data; however, we can pass the `test_size` parameter to change this to any other ratio. We generally want to train on as much data as possible but still keep a meaningful amount of unseen data in reserve, and so *75/25* is a commonly accepted ratio in the industry. With this concept in mind, let's move on to developing a full ML model.

Developing an ML model

There are numerous ML models that we interact with on a daily basis as end users, and we likely do not even realize it. Think back to all the activities you did today: scrolling through social media, checking your email, or perhaps you visited a store or a supermarket. In each of these settings, you likely interacted with an already deployed ML model. On social media, the posts that are presented on your feed are likely the output of a supervised **recommendation** model. The emails you opened were likely filtered for spam emails using a **classification** model. And, finally, the number of goods available within the grocery store was likely the output of a **regression** model, allowing them to predict today's demand. In each of these models, a great deal of time and effort was dedicated to ensuring they function and operate correctly. In these situations, while the development of the model is important, the most important thing is how the data is prepared ahead of time. As scientists, we always have a tendency to organize our thoughts and processes as best we can, so let's organize a workflow for the process of developing ML models:

1. **Data acquisition**: Collecting data via SQL queries, local imports, or API requests
2. **EDA and preprocessing**: Understanding and cleaning up the dataset
3. **Model development and validation**: Training a model and verifying the results
4. **Deployment**: Making your model available to end users

With these steps in mind, let's go ahead and develop our first model.

We begin by importing our data. We will use a new dataset that we have not worked with yet known as the `Breast Cancer Wisconsin` dataset. This is a *multivariate* dataset, published in 1995, containing several hundred instances of breast cancer masses. These masses are described in the form of measurements that we will use as *features* (X). The dataset also includes information regarding the malignancy of each of the instances, which we will use for our output *label* ($\hat{y}$). Given that we have both the input data and the output data, this calls for the use of a *classification* model.

Data acquisition

Let's import our data and check its overall shape:

```
import pandas as pd
import numpy as np
df = pd.read_csv("../../datasets/dataset_wisc_sd.csv")
df.shape
```

We notice that there are 569 rows (which we generally call *observations*) and 32 columns (which we generally call *features*) of data. We generally want our dataset to have many more *observations* than *features*. There is no golden rule about the ideal ratio between the two, but you generally want to have at least 10x more observations than features. So, with 32 columns, you would want to have at least 320 observations – which we do in this case!

Exploratory data analysis and preprocessing:

Exploratory Data Analysis (EDA) is arguably one of the most important and time-consuming steps in any given ML project. This step generally consists of many smaller steps whose objectives are as follows:

- Understand the data and its features.
- Address any inconsistencies or missing values.
- Check for any correlations between the features.

Please note that the order in which we carry out these steps may be different depending on your dataset. With all this in mind, let's get started!

Examining the dataset

One of the first steps after importing your dataset is to quickly check the quality of the data. Recall that we can use square brackets ([]) to specify the columns of interest, and we can use the head() or tail() functions to see the first or last five rows of data:

```
df[["id", "diagnosis", "radius_mean", "texture_mean", "concave
points_worst"]].head()
```

We can see the results of this code in *Figure 5.9*:

	id	diagnosis	radius_mean	texture_mean	concave points_worst
0	842302	M	17.99	10.38	0.2654
1	842517	M	20.57	17.77	0.186
2	84300903	M	19.69	21.25	0.243
3	84348301	M	11.42	20.38	0.2575
4	84358402	M	20.29	14.34	0.1625

Figure 5.9 – A sample of the Breast Cancer Wisconsin dataset

We can quickly get a sense of the fact that the data is very well organized, and from a first glance, it does not appear to have any problematic values, such as unusual characters or missing values. Looking over these select columns, we notice that there is a unique identifier in the beginning consisting of *integers*, followed by the diagnosis (**M = malignant** and **B = benign**) consisting of *strings*. The rest of the columns are all features, and they all appear to be of the *float* (decimals) data type. I encourage you to expand the scope of the preceding table and explore all the other features within this dataset.

In addition to exploring the values, we can also explore some of the summary statistics provided by the `describe()` function in the `pandas` library. Using this function, we get a sense of the total count, as well as some descriptive statistics such as the mean, maximum, and minimum values:

```
df[["id", "diagnosis", "radius_mean", "texture_mean",
"perimeter_mean", "area_mean", "concave points_worst"]].
describe()
```

The output of this function can be seen in the following screenshot:

	id	radius_mean	texture_mean	perimeter_mean	area_mean
count	5.690000e+02	569.000000	569.000000	569.000000	567.000000
mean	3.037183e+07	14.127292	19.289649	91.969033	655.657848
std	1.250206e+08	3.524049	4.301036	24.298981	352.288768
min	8.670000e+03	6.981000	9.710000	43.790000	143.500000
25%	8.692180e+05	11.700000	16.170000	75.170000	420.300000
50%	9.060240e+05	13.370000	18.840000	86.240000	551.700000
75%	8.813129e+06	15.780000	21.800000	104.100000	785.600000
max	9.113205e+08	28.110000	39.280000	188.500000	2501.000000

Figure 5.10 – A table of some summary statistics for a DataFrame

Looking over the code, we notice that we had requested the statistics for seven columns, however, only five appeared in the table. We can see that the id values (which are *primary keys* or *unique identifiers*) were summarized here. These values are meaningless, as the mean, maximum, and minimum of a set of primary keys tell us nothing. We can ignore this column for now. We also asked for the diagnosis column; however, the diagnosis column does not use a numerical value. Instead, it contains *strings*. Finally, we see that the concave points_worst feature was also not included in this table, indicating that the data type is not numerical for whatever reason. We will take a closer look at this in the next section when we clean the data.

Cleaning up values

Getting the values within your dataset cleaned up is one of the most important steps when handling an ML project. A famous saying among data scientists when describing models is *garbage in, garbage out*. If you want to have a strong predictive model, then ensuring the data that supports it is of good quality is an important first step.

To begin, let's take a closer look at the data types, given that there may be some inconsistencies here. We can get a sense of the data types for each of the 32 columns using the following code:

```
df.dtypes
```

We can see the output of this code in *Figure 5.11*, where the column names are shown with their respective data types:

id	int64	compactness_se	float64
diagnosis	object	concavity_se	float64
radius_mean	float64	concave points_se	float64
texture_mean	float64	symmetry_se	float64
perimeter_mean	float64	fractal_dimension_se	float64
area_mean	float64	radius_worst	float64
smoothness_mean	float64	texture_worst	float64
compactness_mean	float64	perimeter_worst	float64
concavity_mean	float64	area_worst	float64
concave points_mean	float64	smoothness_worst	float64
symmetry_mean	float64	compactness_worst	float64
fractal_dimension_mean	float64	concavity_worst	float64
radius_se	float64	concave points_worst	object
texture_se	float64	symmetry_worst	float64
perimeter_se	float64	fractal_dimension_worst	float64
area_se	float64	dtype: object	
smoothness_se	float64		

Figure 5.11 – A list of all of the columns in a dataset with their respective data types

Looking over the listed data types, we see that the `id` column is listed as an integer and the `diagnosis` column is listed as an object, which seems consistent with the fact that it appeared to be a single-letter string in *Figure 5.9*. Looking over the features, they are all listed as floats, completely consistent with what we previously saw, with the exception of one feature: `concave points_worst`. This feature is listed as an object, indicating that it might be a string. We noted earlier that this column consisted of float values, and so the column itself should be of the float type. Let's take a look at this inconsistency sooner rather than later. We can make an attempt to *cast* the column to be of the float type instead of using the `astype()` function:

```
df['concave points_worst'] = df['concave points_worst'].
astype(float)
```

However, you will find that this code will error out, indicating that there is a row in which the `\\n` characters are present and it is unable to convert the string to a float. This is known as a *newline character* and it is one of the most common items or *impurities* you will deal with when handling datasets. Let's move on and identify the lines in which this character is present and decide how to deal with it. We can use the `contains()` function to find all instances of a particular string:

```
df[df['concave points_worst'].str.contains(r"\\n")]
```

The output of this function shows that only the row with the `146` index contains this character. Let's take a closer look at the specific cell from the `146` row:

```
df["concave points_worst"].iloc[146]
```

We see that the cell contains the `0.1865\\n\\n` string. It appears as though the character is printed twice and only in this row. We could easily open the CSV file and correct this value manually, given that it only occurred a single time. However, what if this string appeared 10 times, or 100 times? Luckily, we can use a **Regular Expression** (**regex**) to *match* values, and we can use the `replace()` function to *replace* them. We can specifically chain this function on `df` instead of the single column to ensure that the function parses the full DataFrame:

```
df = df.replace(r'\\n','', regex=True)
```

A regex is a powerful tool that you will often rely on for various text matching and cleaning tasks. You can remove spaces, numbers, characters, or unusual combinations of characters using regex functions. We can double-check the regex function's success by once again examining that specific cell's value:

```
df["concave points_worst"].iloc[146]
```

The value is now only `0.1865`, indicating that the function was, in fact, successful. We can now cast the column's type to a float using the `astype()` function and then also confirm the correct datatypes are listed using `df.dtypes`.

So far, we were able to address issues in which invalid characters made their way into the dataset. However, what about items that are missing? We can run a quick check on our dataset to determine if any values are missing using the `isna()` function:

```
df.isna().values.sum()
```

The value returned shows that there are seven rows of data in which a value is missing. Recall that in *Chapter 4, Visualizing Data with Python*, we looked over a few methods to address missing values. Given that we have a sufficiently large dataset, it would be appropriate to simply eliminate these few rows using the `dropna()` function:

```
df = df.dropna()
```

We can check the dataset's shape before and after implementing the function to ensure the proper number of rows was in fact dropped.

Taking some time to clean up your dataset ahead of time is always recommended, as it will help prevent problems and unusual errors down the line. It's always important to check the *data types* and *missing values*.

Understanding the meaning of the data

Let's now take a closer look at some of the data within this dataset, beginning with the output values in the second column. We know these values correspond to the labels as being *M* for *malignant* and *B* for *benign*. We can use the `value_counts()` function to determine the sum of each category:

```
df['diagnosis'].value_counts()
```

The results show that there are 354 instances of benign masses and 208 instances of malignant masses. We can visualize this ratio using the `seaborn` library:

```
sns.countplot(df['diagnosis'])
```

The output of this code can be seen as follows:

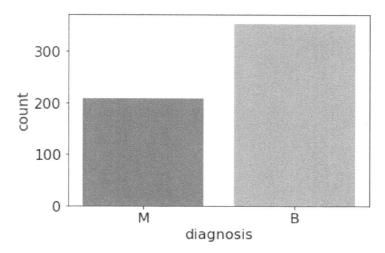

Figure 5.12 – A bar plot showing the number of instances for each class

In most ML models, we try to ensure that the output column is *well balanced*, in the sense that the categories are *roughly equal*. Training a model on an imbalanced dataset with, for example, 95 rows of malignant observations and 5 rows of benign observations would lead to an imbalanced model with poor performance. In addition to visualizing the diagnosis or output column, we can also visualize the features to get a sense of any trends or correlations using the pairplot() function we reviewed in *Chapter 4, Visualizing Data with Python*. We can implement this with a handful of features:

```
sns.pairplot(df[["diagnosis", "radius_mean", "concave points_
mean", "texture_mean"]], hue = 'diagnosis')
```

The following graph shows the output of this:

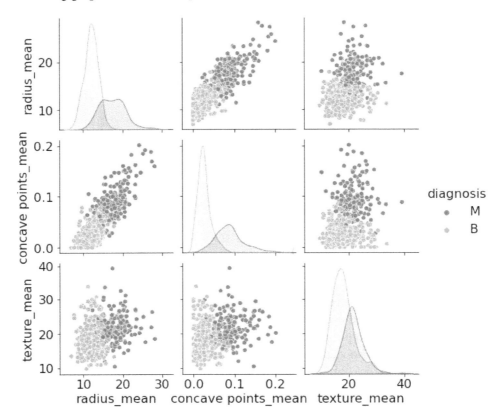

Figure 5.13 – A pair plot of selected features

Looking over these last few plots, we notice a distinguishable separation between the two clusters of data. The clusters appear to exhibit some of the characteristics of a **normal distribution**, in the sense that most points are localized closer to the center, with fewer points further away. Given this nature, one of the first models we may try within this dataset is a **Naïve Bayes classifier**, which tends to work well for this type of data. However, we will discuss this model in greater detail later on in this chapter.

In each of these plots, we see some degree of overlap between the two classes, indicating that two columns alone are not enough to maintain a good degree of separation. So, we could ensure that our ML models utilize more columns or we could try to eliminate any potential outliers that may contribute to this overlap – or we could do both!

First, we can utilize some descriptive *statistics*. Specifically, we can use the **Interquartile Range (IQR)** to identify any potential outliers. Let's examine this for the malignant masses. We begin by isolating the malignant observations in their own variable called dfm. We can then define the first quartile (Q1) and third quartile (Q3) using the radius_mean feature:

```
dfm = df[df["diagnosis"] == "M"]
Q1 = dfm['radius_mean'].quantile(0.25)
Q3 = dfm['radius_mean'].quantile(0.75)
IQR = Q3 - Q1
```

We can then print the outputs of these variables to determine the IQR in conjunction with the mean() and median() functions to get a sense of the distribution of the data. We can visualize these metrics alongside the upper and lower ranges using the boxplot() function provided in seaborn:

```
sns.boxplot(x='diagnosis', y='radius_mean', data=df)
```

This gives us *Figure 5.14*:

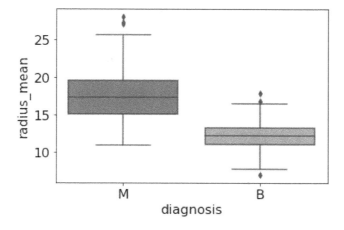

Figure 5.14 – A box-whisker plot of the radius_mean feature

Using the upper and lower ranges, we can filter the DataFrame to exclude any data that falls outside of this scope using the `query()` class within the `pandas` library:

```
df = df.query('(@Q1 - 1.5 * @IQR) <= radius_mean <= (@Q3 + 1.5
* @IQR)')
```

With the code executed, we have successfully removed several outliers from our dataset. If we go ahead and replot the data using one of the preceding scatter plots, we will see that while some of the overlap was indeed reduced, there is still considerable overlap between the two classes, indicating that any future models we develop will need to take advantage of multiple columns to ensure adequate separation as we begin developing a robust classifier. Before we can start training any classifiers, we will first need to address any potential *correlations* within the features.

Finding correlations

With the outliers filtered out, we are now ready to start taking a closer look at any correlations between the features in our dataset. Given that this dataset consists of 30 features, we can take advantage of the `corr()` class we implemented in *Chapter 4, Visualizing Data with Python*. We can create a **heat map** visual by using the `corr()` function and the `heatmap()` function from `seaborn`:

```
f, ax=plt.subplots( figsize = (20,15))
sns.heatmap(df.corr(), annot= True, fmt = ".1f", ax=ax)
plt.xticks(fontsize=18)
plt.yticks(fontsize=18)
plt.title('Breast Cancer Correlation Map', fontsize=18)
plt.show()
```

The output of this code can be seen in *Figure 5.15*, showing a heat map of the various features in which the most correlated features are shown in a lighter color and the least correlated features are shown in a darker color:

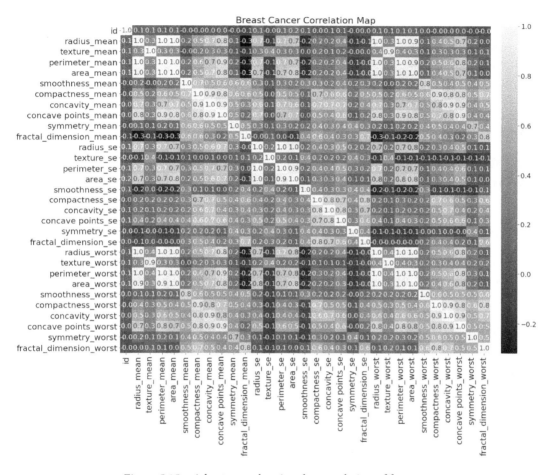

Figure 5.15 – A heat map showing the correlation of features

As we look at this heat map, we see that there is a great deal of correlation between multiple features within this dataset. Take, for instance, the very strong correlation between the `radius_worst` feature and the `perimeter_mean`, and `area_mean` features. When there are strong correlations between independent variables or features within a dataset, this is known as **multicollinearity**. From a statistical perspective, these correlations within an ML model can lead to less reliable statistical inferences, and therefore, less reliable results. To ensure that our dataset is purged of any potential problems of this nature, we simply drop the columns that present a very high degree of correlation with any others. We can identify the correlations using the `corr()` function and create a matrix of these values. We can then select the upper triangle (half of the heat map) and then identify the features whose correlations are greater than `0.90`:

```
corr_matrix = df.corr().abs()
upper = corr_matrix.where(np.triu(np.ones(corr_matrix.shape),
k=1).astype(np.bool))
to_drop = [column for column in upper.columns if
any(upper[column] > 0.90)]
```

The `to_drop` variable now represents a list of columns that should be dropped to ensure that any correlations above the threshold we set are effectively removed. Notice that we used **list comprehension** (a concept that we talked about in *Chapter 2, Introducing Python and the Command Line*) to iterate through these values quickly and effectively. We can then go ahead and drop the columns from our dataset:

```
df.drop(to_drop, axis=1, inplace=True)
```

We can once again plot the heat map to ensure that any potential collinearity is addressed:

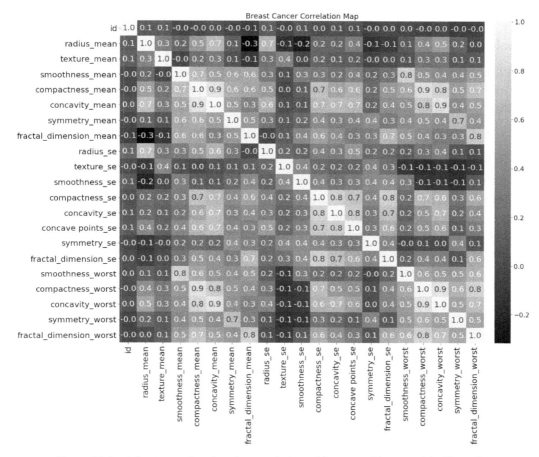

Figure 5.16 – A heat map showing the correlation of features without multicollinearity

Notice that the groups of highly correlated features are no longer present. With the correlations now addressed, we have not only ensured that any potential models we created will not suffer from any performance problems relating to *multicollinearity*, but we also inadvertently reduced the size of the dataset from 30 columns of features down to only 19, making it a little easier to handle and visualize! With the dataset now completely preprocessed, we are now ready to start training and preparing some ML models.

Developing and validating models

Now that the data is ready to go, we can explore a few models. Recall that our objective here is to develop a *classification* model. Therefore, our first step will be to separate our X and $\hat{y}$ values.

1. We will create a variable, X, representing all of the features within the dataset (excluding the id and diagnosis columns, as these are not features). We will then create a variable, y, representing the output column:

    ```
    X = df.drop(columns = ["id", "diagnosis"])
    y = df.diagnosis.values
    ```

    Within most of the datasets we will work with, we will generally see a large difference in the *magnitude* of values, in the sense that one column could be on the order of 1,000, and another column could be on the order of 0.1. This means that features with far greater values will be perceived by the model to make far greater contributions to a prediction – which is not true. For example, think of a project in which we are trying to predict the lipophilicity of a molecule using 30 different features, with one of those being the molecular weight – a feature with a significantly large value but not that large a contribution.

2. To address this challenge, values within a dataset must be **normalized** (or **scaled**) through various functions – for example, the StandardScaler() function from the sklearn library:

    ```
    from sklearn.preprocessing import StandardScaler
    scaler = StandardScaler()
    X_scaled = pd.DataFrame(scaler.fit_transform(X), columns
    = X.columns)
    ```

3. With the features now normalized, our next step is to split the data up into our *training* and *testing* sets. Recall that the purpose of the training set is to train the model, and the testing set is to test the model. This is done to avoid any *overfitting* in the development process:

    ```
    from sklearn.model_selection import train_test_split
    X_train, X_test, y_train, y_test = train_test_split(X_
    scaled, y, random_state=40)
    ```

With the data now split up into four variables, we are now ready to train a few models, beginning with the **Gaussian Naïve Bayes classifier**. This model is a supervised algorithm based on the application of Bayes' theorem. The model is called *naïve* because it makes the assumption that the *features* of each *observation* are independent of one another, which is rarely true. However, this model tends to show strong performance anyway. The main idea behind the Gaussian Naïve Bayes classifier can be examined from a *probability* perspective. To explain what we mean by this, consider the following equation:

$$p(\ label \mid data\ ) = \frac{p(\ data \mid label\ ) * p(\ label\ )}{p(\ data\ )}$$

This states that the probability of the label (given some data) is equal to the probability of the data (given a label – Gaussian, given the normal distribution) multiplied by the probability of the label (prior probability), all divided by the probability of the data (predictor prior probability). Given the simplicity of such a model, Naïve Bayes classifiers can be extremely fast to use in relation to more complex models.

4. Let's take a look at its implementation. We will begin by importing our libraries of interest:

```
from sklearn.naive_bayes import GaussianNB
from sklearn.metrics import accuracy_score
```

5. Next, we can create an instance of the actual model in the form of a variable we call gnb_clf:

```
gnb_clf = GaussianNB()
```

6. We can then fit or train the model using the training dataset we split off earlier:

```
gnb_clf.fit(X_train, y_train)
```

7. Finally, we can use the trained model to make predictions on the testing data and compare the results with the known values. We can use a simple accuracy score to test the model:

```
gnb_pred = gnb_clf.predict(X_test)
    print(accuracy_score(gnb_pred, y_test))
    0.95035
```

With that, we have successfully developed a model performing with roughly 95% accuracy – not a bad start for our first model!

8. While accuracy is always a fantastic metric, it is not the only metric we can use to assess the performance of a model. We can also use **precision**, **recall**, and **f1 scores**. We can quickly calculate these and get a better sense of the model's performance using the `classification_report()` function provided by `sklearn`:

```
from sklearn.metrics import classification_report
print(classification_report(gnb_pred, y_test))
```

Looking at the following output, we can see our two classes of interest (B and M) listed with their respective metrics: `precision`, `recall`, and `f1-score`:

```
              precision    recall  f1-score   support

           B       0.96      0.97      0.96        94
           M       0.93      0.91      0.92        47

    accuracy                           0.95       141
   macro avg       0.95      0.94      0.94       141
weighted avg       0.95      0.95      0.95       141
```

Figure 5.17 – The classification report of the Naïve Bayes classifier

We will discuss these metrics in much more detail in *Chapter 7, Understanding Supervised Machine Learning*. For now, we can see that all of these metrics are quite high, indicating that the model performed reasonably well.

Saving a model for deployment

When an ML model has been trained and is operating at a reasonable level of accuracy, we may wish to make this model available for others to use. However, we would not directly deliver the data or the code to data engineers to deploy the model into production. Instead, we would want to deliver a single trained model that they can take and deploy without having to worry about any moving pieces. Luckily for us, there is a great library known as `pickle` that can help us *gather* the model into a single entity, allowing us to *save* the model. Recall that we explored the `pickle` library in *Chapter 2, Getting Started with Python and the Command Line*. We *pickle* a model, such as the model we named `gnb_clf`, by using the `dump()` function:

```
import pickle
pickle.dump(gnb_clf, open("../../models/gnb_clf.pickle", 'wb'))
```

To prove that the model did in fact save correctly, we can load it using the `load()` function, and once again, we can calculate the accuracy score:

```
loaded_gnb_clf = pickle.load(open("../../models/gnb_clf.
pickle", 'rb'))
loaded_gnb_clf.score(X_test, y_test)
```

Notice that the output of this scoring calculation results in the same value (95%) as we saw earlier, indicating that the model did, in fact, save correctly!

Summary

In this chapter, we took an ambitious step toward understanding some of the most important and useful concepts in ML. We looked over the various terms used to describe the field as it relates to the domain of AI, examined the main areas of ML and the governing categories of *supervised* and *unsupervised* learning, and then proceeded to explore the full process of developing an ML model for a given dataset.

While developing our model, we explored many useful steps. We explored and preprocessed the data to remove inconsistencies and missing values. We also examined the data in great detail, and we subsequently addressed issues relating to *multicollinearity*. Next, we developed a *Gaussian Naïve Bayes* classification model, which operated with a robust 95% rate of accuracy – on our first try too! Finally, we looked at one of the most common ways data scientists hand over their fully trained models to data engineers to move ML models into production.

Although we took the time within this chapter to understand ML within the scope of a supervised classifier, in the following chapter, we will gain a much better understanding of the nuances and differences as we train several unsupervised models.

6
Unsupervised Machine Learning

Oftentimes, many data science tutorials that you will encounter in courses and training revolve around the field of **Supervised Machine Learning (SML)** in which data and its corresponding labels are used to develop predictive models to automate tasks. However, in real-world data, the availability of pre-labeled or categorized data is seldom the case, and most datasets you will encounter will be in their raw and unlabeled form. For cases such as these, or whose primary objectives are more exploratory or not necessarily of automatable fashion, the field of unsupervised ML will be of great value.

Over the course of this chapter, we will explore many methods relating to the areas of clustering and **Dimensionality Reduction (DR)**. The main topics we will explore are listed here:

- Introduction to **Unsupervised Learning (UL)**
- Understanding clustering algorithms
- Tutorial – breast cancer prediction via clustering
- Understanding DR
- Tutorial – exploring DR models

With these topics in mind, let's now go ahead and get started!

Introduction to UL

We will define UL as a subset of ML in which models are trained without the existence of categories or labels. Unlike its supervised counterpart, UL relies on the development of models to capture patterns in the form of features to extract insights from the data. Let's now take a closer look at the two main categories of UL.

There exist many different methods and techniques that fall within the scope of UL. We can group these methods into two main categories: those with **discrete** data (**clustering**) and those with **continuous** data (**DR**). We can see a graphical representation of this here:

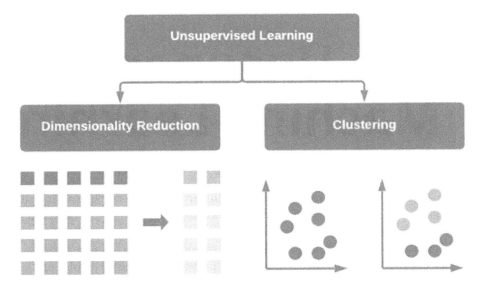

Figure 6.1 – The two types of UL

In each of these techniques, data is either grouped or transformed in order to determine labels or extract insights and representations without knowing the labels or categories of the data ahead of time. Take, for example, the breast cancer dataset we worked with in *Chapter 5, Understanding Machine Learning*, in which we developed a classification model. We trained the model by explicitly telling it which observations within the data were malignant and which were benign, thus allowing it to learn the differences through the features. Similar to our supervised model, we can train an unsupervised **clustering** model to make similar predictions by clustering our data into groups (malignant and benign) without knowing the labels or classes ahead of time. There are many different types of clustering models we can use, and we will explore a few of these in the following section, and others further along in this chapter.

In addition to clustering our data, we can also explore and transform our data through a method known as **DR**, which we will define as the transformation of high-dimensional data into a lower-dimensional space in which the meaningful properties of the features are retained. Data transformations can either be used to reduce the number of features down to a few or to engineer new and useful features for a given dataset. One of the most popular methods that fall within this category is a process known as **Principal Component Analysis (PCA)**—we will explore this specific model in detail further along in this chapter.

Within the scope of both of these categories falls a niche field that is not quite yet a third category given its broad application—this is known as **anomaly detection**. Anomaly detection within the scope of UL, as the name suggests, is a method for the detection of anomalies within an unlabeled dataset. Note that, unlike clustering methods in which there is generally a balance within the different labels of a dataset (for example, 50:50), anomalies tend to be rare in the sense that the number of observations is usually anything but balanced. The most popular methods today when it comes to anomaly detection from an unsupervised perspective tend to not only include **clustering** and **DR**, but also **neural networks** and **isolation forests**.

Now that we've gained a sense of some of the high-level concepts relating to UL and know our objectives, let's now go ahead and get started with some details and examples for each.

Understanding clustering algorithms

One of the most common methods that fall within the category of UL is **clustering analysis**. The main idea behind clustering analysis is the grouping of data into two or more categories of a similar nature to form groups or **clusters**. Within this section, we will explore these different clustering models, and subsequently apply our knowledge in a real-world scenario concerning the development of predictive models for the detection of breast cancer. Let's go ahead and explore some of the most common clustering algorithms.

Exploring the different clustering algorithms

There exists not one, but a broad spectrum of clustering algorithms, each with its own approach to how to best cluster data depending on the dataset at hand. We can divide these clustering algorithms into two general categories: **hierarchical** and **partitional** clustering. We can see a graphical representation of this here:

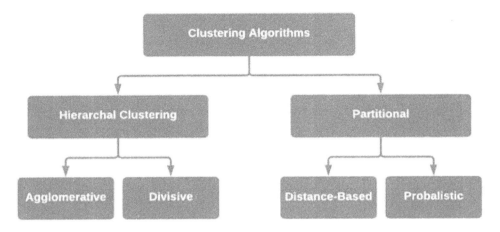

Figure 6.2 – The two types of clustering algorithms

With these different areas of clustering in mind, let's now go ahead and explore these in more detail, beginning with hierarchical clustering.

Hierarchical clustering

Hierarchical clustering, as the name suggests, is a method that attempts to cluster data based on a given hierarchy using two types of approaches: **agglomerative** or **divisive**. Agglomerative clustering is known as a *bottom-up* approach in which each observation in a dataset is assigned its own cluster and is subsequently merged with other clusters to form a hierarchy. Alternatively, **divisive clustering** is a *top-down* approach in which all observations for a given dataset begin in a single cluster and are then split up. We can see a graphical representation of this here:

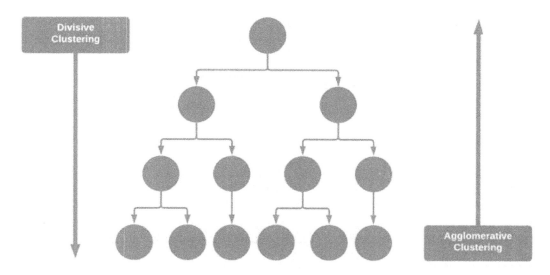

Figure 6.3 – The difference between agglomerative and divisive clustering

With the concept of hierarchical clustering in mind, we can imagine a number of useful applications this can help us with when it comes to phylogenetic trees and other areas of biology. On the other hand, there also exist other methods of clustering in which hierarchy is not accounted for, such as when using **Euclidean** distance.

Euclidean distance

In addition to hierarchical clustering, we also have a set of models that fall under the idea of **partition-based clustering**. The main idea here is separating or partitioning your dataset to form clusters using a given method. Two of the most common types of partition-based clustering are **distance-based clustering** and **probability-based clustering**. When it comes to distance-based clustering, the main idea here is determining whether a given data point belongs to a cluster based solely on distance such as **Euclidean distance**. An example of this is the **K-Means** clustering algorithm—one of the most common clustering algorithms, given its simplicity.

Note that **Euclidean** distance, sometimes referred to as **Pythagorean** distance, from a mathematical perspective, is defined as the distance between two points on a Cartesian coordinate system. For example, for two points, p ($p1$, $p2$) and q ($q1$, $q2$), the Euclidean distance can be calculated as follows:

$$d(p,q) = \sqrt{(q_1 - p_1)^2 + (q_2 - p_2)^2}$$

Within the context of two dimensions, this model is fairly simple and easy to calculate. However, the complexity of this model can increase when given more dimensions, simply represented as follows:

$$d(p,q) = \sqrt{(p_1 - q_1)^2 + (p_2 - q_2)^2 + \cdots + (p_n - q_n)^2}$$

Now that we have gained a better sense of the concept of Euclidean distance, let's now take a look at an actual application known as K-Means.

K-Means clustering

With the concept of Euclidean distance in mind, let's now take a close look at how this can be applied within the context of K-Means. The K-Means algorithm attempts to cluster data by separating samples into k groups consisting of equal variance and minimizing a **criterion** (inertia). The algorithm's objective is to select k **centroids** that minimize the inertia.

The K-Means model is quite simple in the sense that it operates in three simple steps, represented as stars in the following diagram:

Figure 6.4 – K-Means clustering steps

First, a specified number of k **centroids** are randomly initialized. Second, each of the observations, represented by the circles, is then clustered based on distance. The mean of all observations in a given cluster is then calculated, and the centroid is moved to that mean. The process repeats over and over until convergence is reached based on a predetermined threshold.

K-Means is one of the most commonly used clustering algorithms out there, given its simplicity and relatively acceptable computation. It works well with high-dimensional data and is relatively easy to implement. However, it does have its limitations in the sense that it does make the assumption that the clusters are of a spherical nature, which often leads to the misgrouping of data with clusters of non-spherical shapes. Take, for example, another dataset in which the clusters are not of a spherical nature but are more ovular. The application of the **K-Means** model, which operates on the notion of **distance**, would not yield the most accurate results, as shown in the following screenshot:

Figure 6.5 – K-Means clustering with non-spherical clusters

When operating with non-spherical clusters, a good alternative to a **distance**-based model would be a statistical-based approach such as a **Gaussian Mixture Model** (**GMM**).

GMMs

GMMs, within the context of clustering, are algorithms that consist of a particular number of **Gaussian distributions**. Each of these distributions represents a particular cluster. So far within the confines of this book, we have not yet discussed Gaussian distributions—a concept you will often hear about and come across throughout your career as a data scientist. Let's go ahead and define this.

A **Gaussian distribution** can be thought of as a statistical equation representing data points that are symmetrically distributed around their mean value. You will often hear this distribution referred to as a bell curve. We can represent the **probability density function** of a Gaussian distribution as follows:

$$f(x \mid \mu, \sigma^2) = \frac{1}{\sqrt{2\pi\sigma^2}} e^{\frac{-(x-\mu)^2}{2\sigma^2}}$$

Here, μ represents the mean and σ^2 represents the variance. Note that this function represents a single variable. Upon the addition of other variables, we would begin to venture into the space of multivariate Gaussian models, in which x and μ represent vectors of length d. In a dataset consisting of k clusters, we would need a mixture of k Gaussian distributions, in which each distribution has a mean and variance. These two values are determined through a technique known as **Expectation-Maximization** (**EM**).

We will define **EM** as an algorithm that determines the proper parameters for a given model when some data is considered missing or incomplete. These missing or incomplete items are known as **latent variables**, and within the confines of UL, we can consider the actual clusters to be unknown. Note that if the clusters were known, we would be able to determine the mean and variance; however, we need to know the mean and variance to determine the cluster (think of the classic chicken-or-egg situation). We can use EM within the scope of the data to determine the proper values of these two variables to best fit the model parameters. With all this in mind, we are now in a position to discuss GMMs more intelligently.

We previously defined a GMM as a model consisting of multiple Gaussian distributions. We will now elaborate on this definition by including the fact that it is a probabilistic model consisting of multiple Gaussian distributions and utilizes a **soft clustering** approach by determining the membership of a data point to a given cluster based on probability rather than a distance. Notice that this is in contrast to K-Means, which utilizes a **hard clustering** approach. Using the previous example dataset shown in *Figure 6.5* in the previous section, the application of a GMM would likely lead to improved results, as depicted here:

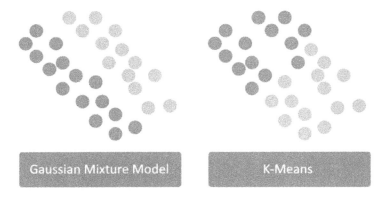

Figure 6.6 – K-Means clustering versus GMMs

Within this section, we discussed a few of the most common clustering algorithms commonly used in many applications within the field of biotechnology. We see clustering being applied in areas such as bio-molecular data, scientific literature, manufacturing, and even oncology, as we will experience in the following tutorial.

Tutorial – breast cancer prediction via clustering

Over the course of this tutorial, we will explore the application of commonly used clustering algorithms for the analysis and prediction of cancer using the `Wisconsin Breast Cancer` dataset we applied in *Chapter 5, Understanding Machine Learning*. When we last visited this dataset, we approached the development of a model from the perspective of a supervised classifier in which we knew the labels of our observations ahead of time. However, in most real-world scenarios, knowledge of the labels ahead of time is rare. **Clustering analysis**, as we will soon see, can be highly valuable in these situations, and can even be used to label data to use within the context of a classifier later on. Over the course of this tutorial, we will develop our models using the data but pretend that we do not know the labels ahead of time. We will only use known labels to compare the results of our models. With this in mind, let's go ahead and get started!

We will begin by importing our dataset as we have previously done and check the shape, as follows:

```
df = pd.read_csv("../../datasets/dataset_wisc_sd.csv")
print(df.shape)
```

We notice that there are 569 rows of data in this dataset. In our previous application, we had cleaned up the data to address missing and corrupt values. Let's go ahead and clean those up, as follows:

```
df = df.replace(r'\\n','', regex=True)
df = df.dropna()
print(df.shape)
```

With the current shape of the data consisting of 569 rows with 32 columns, this now matches our previous dataset, and we are now ready to proceed.

Although we will not be using these labels to develop any models, let's take a quick look at them, as follows:

```
import seaborn as sns
sns.countplot(df['diagnosis']);
```

We can see in the following screenshot that there are two classes—M for malignant and B for benign. The two classes are not perfectly balanced but will do for the purposes of our clustering model:

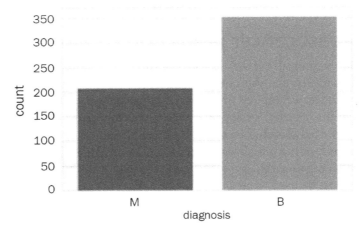

Figure 6.7 – The distribution of the two classes

To make our comparison to these labels easier during the following steps of our clustering analysis, let's go ahead and encode these labels as numerical values in which we will convert M to 1 and B to 0, as follows:

```
df['diagnosis'] = df['diagnosis'].map({'M':1,'B':0})
```

We can use the df.head() function to see the first few rows of our dataset and confirm that the diagnosis column did in fact get encoded properly. Next, we will prepare a quick pairplot of a few select features, as follows:

```
select_feats = ["diagnosis", "radius_mean", "texture_mean",
"smoothness_mean"]
sns.pairplot(df[select_feats], hue = 'diagnosis', markers=["s",
"o"])
```

We can use the markers argument to specify two distinct shapes to plot the two classes, yielding the following pairplot showing scatter plots of our features:

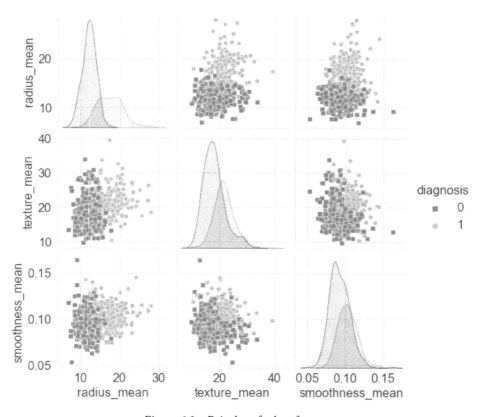

Figure 6.8 – Pairplot of select features

Our first objective is to look over the many features and get a sense of which two features show the least amount of overlap or the best degree of separation. We can see that the smoothness_mean and texture_mean columns have a high degree of overlap; however, radius_mean and texture_mean seem less so. We can take a closer look at these by plotting a scatter plot using the seaborn library, as follows:

```
sns.scatterplot(x="radius_mean", y="texture_mean",
hue="diagnosis", style='diagnosis', data=df, markers=["s",
"o"])
```

Notice that once again, we can use the `style` and `markers` arguments to shape the data points, thus yielding the following diagram as output:

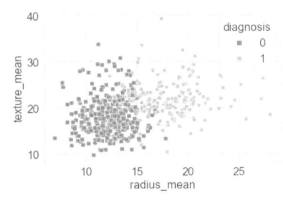

Figure 6.9 – Scatter plot of the two features that showed good separation

Next, we will normalize our data. In statistics, normalization or standardization can have a wide variety of meanings and are sometimes used interchangeably. We will define normalization to mean the rescaling of values into a range of [0,1]. On the other hand, we will define standardization to mean the rescaling of data to have a mean value of 0, and a standard deviation value of 1. For the purposes of our current objectives, we will want to standardize our data as we have previously done using the `StandardScaler` class. Recall that this class standardizes features within the dataset by removing the mean and scaling to variance, which can be represented as follows:

$$z = \frac{(x - u)}{s}$$

Here, x is the standard score of a sample, u is the mean, and s is the standard deviation. We can apply this in Python with the following code:

```
from sklearn.preprocessing import StandardScaler
scaler = StandardScaler()
```

```
X = df.drop(columns = ["id", "diagnosis"])
y = df.diagnosis.values
X_scaled = pd.DataFrame(scaler.fit_transform(X), columns =
X.columns)
```

With our dataset scaled, we are now ready to start applying a few models. We will begin with the agglomerative clustering model from the `sklearn` library.

Agglomerative clustering

Recall that **agglomerative** clustering is a method in which clusters are formed by recursively merging clusters together. Let's go ahead and implement the agglomerative clustering algorithm with our dataset, as follows:

1. First, we will import the specific class of interest from the `sklearn` library, and then create an instance of our model by specifying the number of classes we want and setting the linkage as `ward`—one of the most common agglomerative clustering methods used. The code is illustrated in the following snippet:

    ```
    from sklearn.cluster import AgglomerativeClustering
    agc = AgglomerativeClustering(n_clusters=2,
    linkage="ward")
    ```

2. Next, we will fit our model to our dataset, and predict the clusters to which they belong. Notice in the following code snippet that we used the `fit_predict()` function, using the first two features, `radius_mean` and `texture_mean`, and not the whole dataset:

    ```
    agc_featAll_pred = agc.fit_predict(X_scaled.iloc[:, :2])
    ```

3. We can then use `matplotlib` and `seaborn` to generate a diagram showing the actual (`true`) results on the left and predicted agglomerative clustering results on the right, as follows:

    ```
    import matplotlib.pyplot as plt
    import seaborn as sns
    plt.figure(figsize=(20, 5))
    plt.subplot(121)
    plt.title("Actual Results")
    ax = sns.scatterplot(x="radius_mean", y="texture_mean",
    hue=y, style=y, data=X_scaled, markers=["s", "o"])
    ```

```
ax.legend(loc="upper right")

plt.subplot(122)
plt.title("Agglomerative Clustering")
ax = sns.scatterplot(x="radius_mean", y="texture_mean",
hue=agc_featAll_pred, style=agc_featAll_pred, data=X_
scaled, markers=["s", "o"])
ax.legend(loc="upper right")
```

Notice in the preceding code snippet the use of the `subplot()` functionality in which the value `122` was used to represent `1` as the total number of rows, `2` as the total number of columns, and `2` as the specific index location of the plot. You can view the output here:

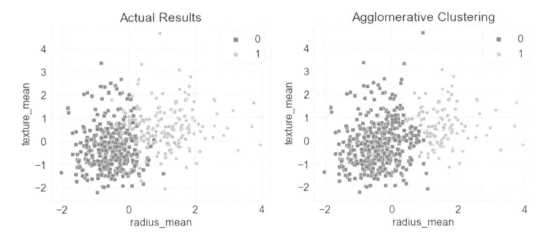

Figure 6.10 – Results of the agglomerative clustering model relative to the actual results

4. From an initial estimation, we see that the model did a fairly reasonable job in distinguishing between the two clusters, having known very little about the actual true outcome. We can get a quick measure of its performance using the `accuracy_score` method from `sklearn`. Although getting a sense of the recall and f-1 scores is also important, we will stick to accuracy for simplicity for now. The code is illustrated in the following snippet:

```
from sklearn.metrics import accuracy_score
print(accuracy_score(y, agc_featAll_pred))
0.832740
```

In summary, the agglomerative clustering model using only the first two features of the dataset yielded an accuracy of roughly 83%—not a bad first attempt! If you are following along using the provided code, I would encourage you to try adding yet another feature and fitting the model with three or four or five features instead of just two and see whether you are able to improve the performance. Better yet, explore the other features provided in this dataset, and see whether you can find others that offer better separation and beat our 83% metric. Let's now investigate the performance of K-Means instead.

K-Means clustering

Let's now investigate the application of **K-Means** clustering using the dataset. Recall that the K-Means algorithm attempts to cluster data by partitioning the data into k clusters based on the location of their centroids. We can apply the K-Means algorithm using the following steps:

1. We will begin by importing the KMeans class from the sklearn library, as follows:

    ```
    from sklearn.cluster import KMeans
    ```

2. Next, we can initialize an instance of the K-Means model and specify the number of clusters being 2, the number of iterations being 10, and the initialization method being k-means++. This initialization setting simply selects the initial cluster centers using an algorithm, with the aim of speeding up convergence. We can adjust the parameters in a process known as tuning in order to maximize the performance of the model. The code is illustrated in the following snippet:

    ```
    kmc = KuMeans(n_clusters=2, n_init=10, init="k-means++")
    ```

3. We can then use the fit_predict() method to fit our data and predict the clusters for each of the observations. Notice in the following code snippet that the model is only fitting and predicting the outcomes based on the first two features alone:

    ```
    kmc_feat2_pred = kmc.fit_predict(X_scaled.iloc[:, :2])
    ```

4. Finally, we can go ahead and plot the results of our predictions in comparison to the true values of the known classes using the seaborn library, as follows:

    ```
    plt.figure(figsize=(20, 5))
    plt.subplot(131)
    plt.title("Actual Results")
    ax = sns.scatterplot(x="radius_mean", y="texture_mean",
    hue=y, style=y, data=X_scaled, markers=["s", "o"])
    ax.legend(loc="upper right")
    ```

```
plt.subplot(132)
plt.title("KMeans Results (Features=2)")
ax = sns.scatterplot(x="radius_mean", y="texture_mean",
hue= kmc_feat2_pred , style= kmc_feat2_pred, data=X_
scaled, markers=["s", "o"])
ax.legend(loc="upper right")
```

Upon executing this code, we get a scatter plot showing our results, as follows:

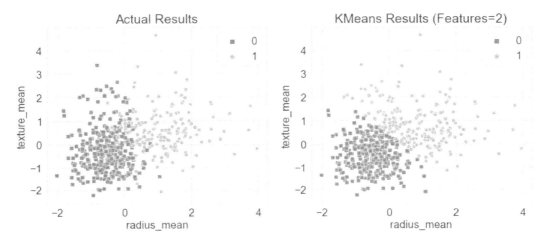

Figure 6.11 – Results of the K-Means clustering model relative to the actual results

While examining the two plots, we notice that the model did a remarkable job at separating the bulk of the data between the two clusters. Notice that **K-Means**, as opposed to agglomerative clustering, separated the boundary between the two clusters quite sharply defined. K-Means is known as a *hard* clustering model in the sense that the centroids and their distance from data points dictate the membership of a data point to a given cluster. Notice that this strongly defined boundary was the case for only the first two features and yielded an accuracy of 86%. Let's go ahead and try this with a few more features.

Although written in a non-Python fashion for illustrative purposes, we can fit out model using 2, 3, 4, and all the features, as follows:

```
kmc_feat2_pred = kmc.fit_predict(X_scaled.iloc[:, :2])
kmc_feat3_pred = kmc.fit_predict(X_scaled.iloc[:, :3])
kmc_feat4_pred = kmc.fit_predict(X_scaled.iloc[:, :4])
kmc_featall_pred = kmc.fit_predict(X_scaled.iloc[:, :])
```

5. Next, using the `subplot()` methodology, we can generate four plots to illustrate the changes in which each individual subplot represents one of the plots depicted. Here's the code we'll need:

```
plt.figure(figsize=(20, 5))
plt.subplot(141)
plt.title("KMeans Results (Features=2)")
ax = sns.scatterplot(x="radius_mean", y="texture_mean",
hue=kmc_feat2_pred, style=kmc_feat2_pred, data=X_scaled,
markers=["s", "o"])
ax.legend(loc="upper right")
# Apply the same for the other plots
```

With the code executed, we yield the following diagram showing the results:

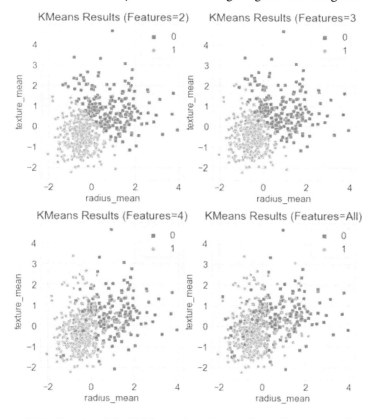

Figure 6.12 – Results of the K-Means clustering model with increasing features

We can calculate the accuracy using only two features to be ~86%, whereas three features yielded 89%. We will notice, however, that the numbers not only begin to plateau with more features included but also decrease when all features were included, yielding a lower accuracy of 82%. Note that as we begin to add more features to the model, we are adding more dimensions. For example, with three features, we are now using a **three-dimensional** (**3D**) model, as shown by the blended border between the two datasets. In some cases, the more features we have, the bigger the strain it will have on a given model. This borders a concept known as the **Curse of Dimensionality** (**COD**) in the sense that the volume of the space begins to increase at an incredible rate given more dimensions, which can impact the performance of the model. We will touch on some of the ways we can remedy this in the future, particularly in the following tutorial, as we begin to discuss **DR**.

In summary, we were able to apply the K-Means model on our dataset and were able to yield a considerable accuracy of 89% using the first three features. Let's now go ahead and explore the application of a statistical method such as GMM.

GMMs

Let's now explore the application of GMMs on our dataset. Recall that these models represent a mixture of probability distributions, and the membership of an observation to a cluster is calculated based on that probability and not on Euclidean distance. With that in mind, let's go ahead and get started, as follows:

1. We can begin by importing the `GaussianMixture` class from the `sklearn` library, like this:

    ```
    from sklearn.mixture import GaussianMixture
    ```

2. Next, we will create an instance of the model and specify the number of components as 2, and set the covariance type as `full` such that each component has its own covariance matrix, as follows:

    ```
    gmm = GaussianMixture(n_components=2, covariance_
    type="full")
    ```

3. We will then fit the data model with our data, once again using only the first two features, and predict the clusters for each of the observations, as follows:

    ```
    gmm_featAll_pred = 1-gmm.fit_predict(X_scaled.iloc[:,
    :2])
    ```

4. Finally, we can go ahead and plot the results using the `seaborn` library, as follows:

```
plt.figure(figsize=(20, 5))
plt.subplot(131)
plt.title("Actual Results")
ax = sns.scatterplot(x="radius_mean", y="texture_mean",
hue=y, style=y, data=X_scaled, markers=["s", "o"])
ax.legend(loc="upper right")
plt.subplot(132)
plt.title("Gaussian Mixture Results (Features=All)")
ax = sns.scatterplot(x="radius_mean", y="texture_mean",
hue=gmm_featAll_pred, style=gmm_featAll_pred, data=X_
scaled, markers=["s", "o"])
ax.legend(loc="upper right")
```

Upon executing our code, we yield the following output, showing the actual results of the dataset relative to our predicted ones:

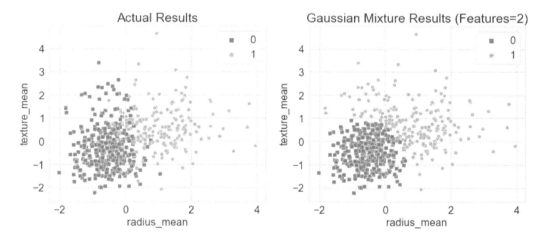

Figure 6.13 – Results of the GMM relative to the actual results

Once again, we can see that the boundary between the two classes is very defined within the Gaussian model, in which there is little to no blending, as the actual results show, thus yielding an accuracy of ~85%. Notice, however, that relative to the K-Means model, the GMM predicted a dense circular distribution in blue, with some members of the orange class wrapping around it in a very non-circular fashion.

Similar to the previous model, we can once again add some more features to this model in an attempt to further improve the performance. However, we see in the following screenshot that despite the addition of more features from left to right, the model does not improve, and the predictive capabilities begin to suffer:

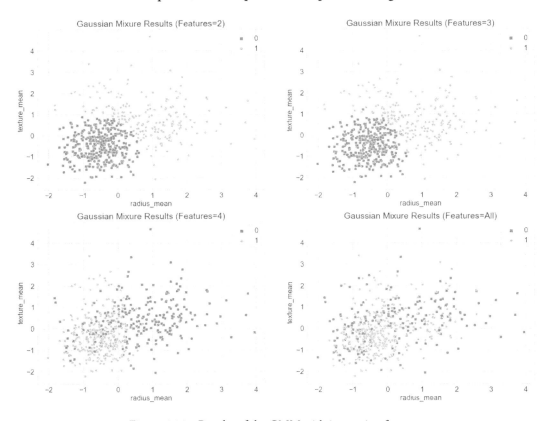

Figure 6.14 – Results of the GMM with increasing features

In summary, over the course of this tutorial, we investigated the use of clustering analysis to develop various predictive models for a dataset while assuming the absence of labels. Throughout the tutorial, we investigated the use of three of the most common clustering models: **agglomerative** clustering, **K-Means** clustering, and GMMs. We investigated the specific properties of all three models and their applicability to the `Wisconsin Breast Cancer` dataset. We determined that the K-Means model using three features showed optimal performance relative to other models, some of which utilized the dataset as a whole. We can speculate that all of the features contribute some level of significance when it comes to predictive power; however, the inclusion of all features within the models showed degraded performance. We will investigate some ways to mitigate this in the following section, pertaining to **DR**.

Understanding DR

The second category of **UL** that we will discuss is known as **DR**. As the full name states, these are simply methods used to reduce the number of dimensions in a given dataset. Take, for example, a highly featured dataset with 100 or so columns—DR algorithms can be used to help reduce the number of columns down to perhaps 5 while preserving the value that each of those original 100 columns contains. You can think of DR as the process of condensing a dataset in a horizontal fashion. The resulting columns can generally be divided into two types: new features, in the sense that a new column with new numerical values was generated in a process known as **Feature Engineering (FE)**, or old features, in the sense that only the most useful columns were preserved in a process known as **feature selection**. Over the course of the following section and within the confines of UL, we will be focusing more on the aspect of FE as we create new features representing reduced versions of many others. We can see a graphical illustration of this concept here:

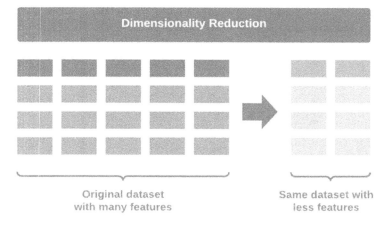

Figure 6.15 – Graphical representation of DR

There are many different methods that we can use to implement DR, each with its own process and underlying theory; however, before we begin implementing these, there is a very important concept we need to address. You are now probably wondering why DR matters. Why would any data scientist eliminate features after another data scientist or data engineer went through all of the trouble to put together a comprehensive and rich dataset to begin with? There are three answers to this question, as outlined here:

- We are not necessarily eliminating any data from our given dataset but are exploring our data from a different window, which may provide some new insights that we would not have seen using the original dataset.

- Developing models with many features is a computationally expensive process, therefore the ability to train our model using fewer features will always be faster, less computationally intensive, and more favorable.

- The use of DR can help reduce noise within the dataset to further improve clustering models and data visualizations.

  With these answers in mind, let's now go ahead and talk about a concept that you will hear in many meetings, discussions, and interviews—the COD.

Avoiding the COD

The COD is regarded as a general phenomenon that arises when handling highly dimensional datasets—a term that was originally coined by Richard E. Bellman. In essence, the COD refers to issues that arise with highly dimensional datasets that do not occur in lower-dimensional datasets of similar size. As the number of features in a given dataset increases, the total number of samples will also increase proportionally. Take, for example, some dataset consisting of one dimension. Within this dataset, let's assume that we would need to examine a total of 10 regions. If we added a second dimension, we would now need to examine a total of 100 regions. Finally, if we added a third dimension, we would now need to examine a total of 1,000 regions. Think back for a moment to some of the datasets we have been working with so far that extend well beyond 1,000 rows and have at least 10 columns—the complexity of datasets such as these can grow quite rapidly. The main takeaway point here is that feature growth has a large impact on the development of a model. We can see a graphical illustration of this here:

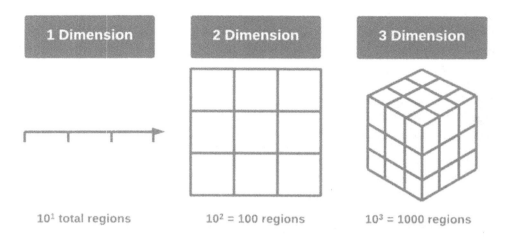

1 Dimension   2 Dimension   3 Dimension

10^1 total regions 10^2 = 100 regions 10^3 = 1000 regions

Figure 6.16 – Graphical representation of the COD

As the number of **features** begins to increase, so does the overall **complexity** of an ML model, which can have a number of negative impacts such as overfitting, thus resulting in poor **performance**. One of the main motivations to reduce the dimensionality of a dataset is to ensure that overfitting is avoided, thus resulting in a more robust model. We can see a graphical illustration of this here:

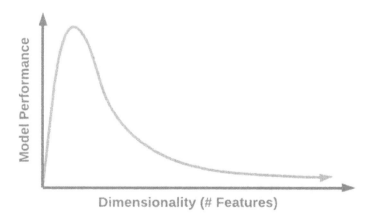

Figure 6.17 – The effect of higher dimensions on model performance

The necessity to reduce datasets from being highly dimensional to a low-dimensional form is especially true in the life science and biotechnology sectors. Throughout the many processes that scientists and engineers face within this field, there are generally hundreds of features relating to any given process. Whether we are looking for datasets relating to protein structures, monoclonal antibody titer, small molecule docking site selection, **Bispecific T-cell Engager (BiTE)** drug design, or even datasets relating to **Natural Language Processing (NLP)**, the reduction of features will always be useful and in many cases necessary for the development of a good ML model.

Now that we have gained a better understanding of DR as it relates to the concept of the COD and the many benefits that can arise from these methods, let's now go ahead and look at a few of the most common models we should know about in this field.

Tutorial – exploring DR models

There are many different ways we can classify the numerous dimensionality algorithms out there, based on type, function, or outcome, and so on. However, for the purposes of getting a strong overview of DR in just a few pages within this chapter, we will classify our models as being either of a linear or non-linear fashion. Linear and non-linear models are two different types of data transformations. We can think of data transformations as methods in which data is altered or reshaped in one way or another. We can loosely define linear methods as transformations in which the output of a model is proportional to its input. Take, for example, p and q being two mathematical vectors.

We can consider a transformation to be linear when the following apply:

- The transformation of p is multiplied by a scalar and its result is the same as multiplying p by the scalar and then applying the transformation.

- The transformation of $p + q$ is the same as the transformation of p + the transformation of q.

If a model does not satisfy these two properties, it is considered a non-linear model. Many different models fall within the scope of these two classes; however, for the purposes of this chapter, we will take a look at four main models that have gained quite a bit of popularity within the data science community over recent years. We can see a graphical illustration of this here:

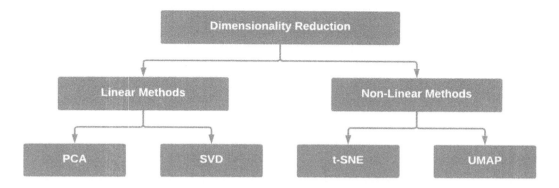

Figure 6.18 – Two examples of models for each of the fields of DR

Within the scope of linear methods, we will take a close look at **PCA** and **Singular Value Decomposition** (**SVD**). In addition, within the scope of non-linear methods, we will take a close look at **t-distributed Stochastic Neighbor Embedding** (**t-SNE**) and **Uniform Manifold Approximation and Projection** (**UMAP**). With these four models in mind and how they fit into the grand scheme of DR, let's go ahead and get started.

PCA

One of the most common and widely discussed forms of UL is **PCA**. PCA is a linear form of DR, allowing users to transform a large dataset of correlated features into a smaller number of uncorrelated features known as principal components. These **principal components**, although numerically fewer than their original features, can still retain as much of the variation or *richness* as the original dataset. We can see a graphical illustration of this here:

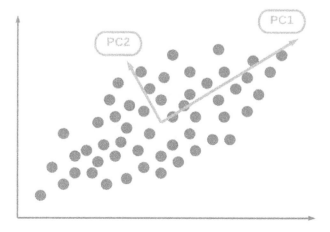

Figure 6.19 – Graphical representation of PCA and its principal components

There are a few things that need to happen in order to effectively implement PCA on any given dataset. Let's take a high-level overview of what these steps are and how they can impact the final outcome. We must first normalize or standardize our data to ensure that the mean is 0 and the standard deviation is 1. Next, we calculate what is known as the **covariance matrix**, which is a square matrix containing the covariance between each of the pairs of elements. In a **two-dimensional** (**2D**) dataset, we can represent a covariance matrix as such:

$$Matrix(Covariance) = \begin{bmatrix} Var[X_1] & Cov[X_1, X_2] \\ Cov[X_2, X_1] & Var[X_2] \end{bmatrix}$$

Next, we can calculate the **eigenvalues** and **eigenvectors** for the covariance matrix, as follows:

$$\det(\lambda I - A) = 0$$

Here, λ is an eigenvalue for a given matrix A, and I is the identity matrix. Using the eigenvector, we can determine the eigenvalue v using the following equation:

$$(\lambda I - A)v = 0$$

Next, the eigenvalues are ordered from the largest to the smallest, which represent the components in order of significance. A dataset with n variables or features will have n eigenvalues and eigenvectors. We can then limit the number of eigenvalues or vectors to a predetermined number, thus reducing the dimensions of our dataset. We can then form what we call a feature vector, using the eigenvectors of interest.

Finally, we can form the **principal components** using the **transpose** of the feature vector, as well as the transpose of the scaled data of the original dataset, and multiplying the two together, as follows:

$$PrincipalComponets = NewFeatureVectors^T * OriginalScaledData^T$$

Here, *PrincipalComponents* is returned as a matrix. Easy, right?

Let's now go ahead and implement PCA using Python, as follows:

1. First, we import PCA from the `sklearn` library and instantiate a new PCA model. We can set the number of components as 2, representing the fact that we only want two components returned to us, and use `full` for `svd_solver`. We can then fit the data on our scaled dataset, as follows:

```
from sklearn.decomposition import PCA
pca_2d = PCA(n_components=2, svd_solver='full')
pca_2d.fit(X_scaled)
```

2. Next, we can transform our data and assign our output matrix to the `data_pca_2d` variable, as follows:

```
data_pca_2d = pca_2d.fit_transform(X_scaled)
```

3. Finally, we can go ahead and plot the results using `seaborn`, as follows:

```
plt.xlabel("Principal Component 1")
plt.ylabel("Principal Component 2")
sns.scatterplot(x=data_pca_2d[:,0], y=data_pca_2d[:,1],
hue=y, style=y, markers=["s", "o"])
```

Upon executing this code, this will yield a scatter plot showing our principal components with our points colored using `y`, as shown here:

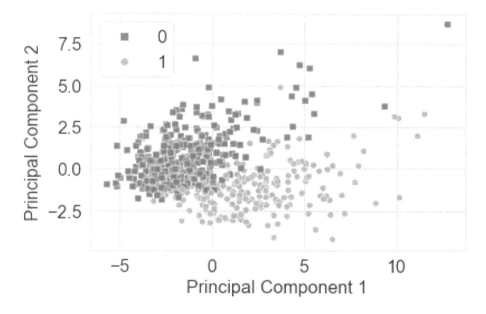

Figure 6.20 – Scatter plot of the PCA results

PCA is a fast and efficient method best used as a precursor to the development of ML models when the number of dimensions has become too complex. Think back for a moment to the dataset we used in our clustering analysis relating to breast cancer predictions. Instead of running our models on the raw or scaled data, we could implement a DR algorithm such as PCA to reduce our dimensions down only two principal components before applying the subsequent clustering model. Remember that PCA is only one of many linear models. Let's now go ahead and explore another popular linear model known as SVD.

SVD

SVD is a popular **matrix decomposition** method commonly used to reduce a dataset to a simpler form. In this section, we will focus specifically on the application of truncated SVD. This model is quite similar to that of PCA; however, the main difference is that the estimator does not center prior to its computation. Essentially, this difference allows the model to be used with sparse matrices quite efficiently.

Let's now introduce and take a look at a new dataset that we can use to apply SVD: *single-cell RNA* (where **RNA** stands for **ribonucleic acid**). The dataset can be found at `http://blood.stemcells.cam.ac.uk/data/nestorowa_corrected_log2_transformed_counts.txt`. This dataset pertains to the topic of single-cell sequencing—a process that examines sequences of individual cells to better understand their properties and functions. Datasets such as these tend to have many columns of data, making them prime candidates for DR models. Let's go ahead and import this dataset, as follows:

```
dfx = pd.read_csv("../../datasets/single_cell_rna/nestorowa_
corrected_log2_transformed_counts.txt", sep=' ',   )
dfx.shape
```

Taking a look at the shape, we can see that there are 3,991 rows of data and 1,645 columns. Relative to the many other datasets we have used, this number is quite large. Within the field of biotechnology, DR is very commonly used to help reduce such datasets into more manageable entities. Notice that the index contains some information about the type of cell we are looking at. To make our visuals more interesting, let's capture this annotation data by executing the following code:

```
dfy = pd.DataFrame()
dfy['annotation'] = dfx.index.str[:4]
dfy['annotation'].value_counts()
```

With the data all set, let's go ahead and implement truncated SVD on this dataset. We can once again begin by instantiating a truncated SVD model and setting the components to 2 with 7 iterations, as follows:

```
from sklearn.decomposition import TruncatedSVD
svd_2d = TruncatedSVD(n_components=2, n_iter=7)
```

Next, we can go ahead and use the `fit_transform()` method to both fit our data and transform the DataFrame to a two-column dataset, as follows:

```
data_svd_2d = svd_2d.fit_transform(dfx)
```

Finally, we can finish things up by plotting our dataset using a scatter plot, and color by annotation. The code is illustrated in the following snippet:

```
sns.scatterplot(x=data_svd_2d[:,0], y=data_svd_2d[:,1],
hue=dfy.annotation, style=dfy.annotation, markers = ["o", "s",
"v"])
```

We can see the results of executing this code in the following screenshot:

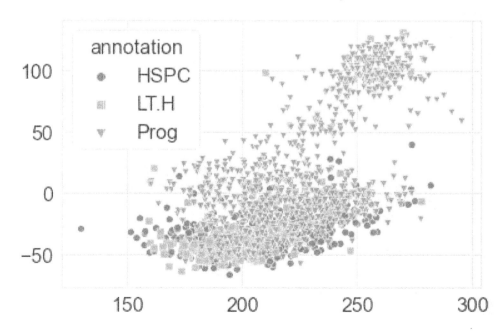

Figure 6.21 – Scatter plot of the results of the SVD model

In the preceding screenshot, we can see the almost 1,400 columns worth of data being reduced to a simple 2D representation—quite fascinating, isn't it? One of the biggest advantages of being able to reduce data in this fashion is that it assists with model development. Let's assume, for the sake of example, that we wish to implement any of our previous clustering algorithms on this extensive dataset. It would take considerably longer to train any given model on a dataset of nearly 1,400 columns compared to a dataset with 2 columns. In fact, if we implemented a GMM on this dataset, the total training time would be **12.4 s ± 158 ms** using the original dataset, relative to **4.06 ms ± 26.6 ms** using the reduced dataset. Although linear models can be very useful when it comes to DR, non-linear models can also be similarly impressive. Next, let's take a look at a popular model known as t-SNE.

t-SNE

On the side of non-linear DR, one of the most popular models commonly seen in action is **t-SNE**. One of the unique features of the t-SNE model relative to the other dimensionality models we have talked about is the fact that it uses probability distribution to represent similarities between neighbors. Simply stated, t-SNE is a statistical method allowing for the DR and visualization of high-dimensional data in which similar points are close together and dissimilar ones are further apart.

t-SNE is a type of **manifold** model, which from a mathematical perspective is a topological space resembling **Euclidean** space. The concept of a manifold is complex, extensive, and well beyond the scope of this book. For the purposes of simplicity, we will state that manifolds describe a large number of geometric surfaces such as a sphere, torus, or cross surface. Within the confines of the t-SNE model, the main objective is to use geometric shapes to give users a feel or intuition of how the high-dimensional data is arranged or organized.

Let's now take a close look at the application of t-SNE using Python. Once again, we can apply this model on our single-cell RNA dataset and get a sense of what the high-dimensional organization of this data looks like from a geometric perspective. Many parameters within t-SNE can be changed and tuned to fit given purposes; however, there is one in particular worth mentioning briefly—perplexity. **Perplexity** is a parameter related to the number of nearest neighbors used as input when it comes to manifold learning. The `scikit-learn` library recommends considering values between 5 and 50. Let's go ahead and take a look at a few examples.

Implementing this model is quite simple, thanks to the high-level **Application Programming Interface (API)** provided by `scikit-learn`. We can begin by importing the `TSNE` class from `scikit-learn` and setting the number of components to 2 and the perplexity to `10`. We can then chain the `fit_transform()` method using our dataset, as illustrated in the following code snippet:

```
from sklearn.manifold import TSNE
data_tsne_2d_p10 = TSNE(n_components=2, perplexity=10.0).fit_
transform(dfx)
```

We can then go ahead and plot our data to visualize the results using `seaborn`, as follows:

```
sns.scatterplot(x=data_tsne_2d_p10[:,0], y=data_tsne_2d_
p10[:,1], hue=dfy.annotation, style=dfy.annotation, markers =
["o", "s", "v"])
```

We can see the output of this in the following screenshot:

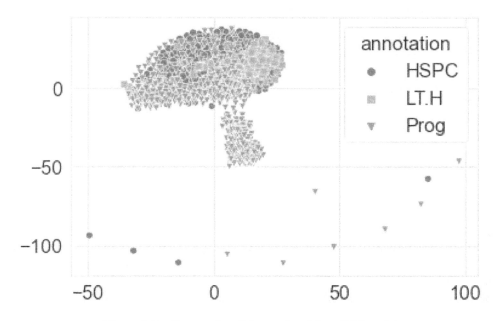

Figure 6.22 – Scatter plot of the results of the t-SNE model

Quite the result! We can see in the preceding screenshot that the model, without any knowledge of the labels, made a 2D projection of the relationship between the data points using the huge dataset it was given. The geometric shape produced gives us a sense of the *look* and *feel* of the data. We can see based on this depiction that a few points seem to be considered outliers as they are depicted much further away, like islands relative to the main continent. Recall that we used a perplexity value of 10 for this particular diagram. Let's go ahead and explore this parameter using a few different values, as follows:

```
data_tsne_2d_p1 = TSNE(n_components=2, perplexity=1.0).fit_
transform(dfx)

data_tsne_2d_p10 = TSNE(n_components=2, perplexity=10.0).fit_
transform(dfx)

data_tsne_2d_p30 = TSNE(n_components=2, perplexity=30.0).fit_
transform(dfx)
```

Using these calculated values, we can visualize them next to each other using the seaborn library, as follows:

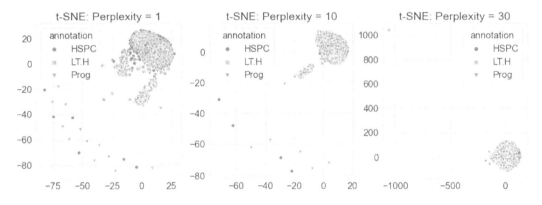

Figure 6.23 – Scatter plots of the t-SNE model with increasing perplexities

When it comes to high-dimensional data, t-SNE is one of the most commonly used models to not only reduce your dimensions but also explore your data by getting a unique feel for its features and their relationships. Although t-SNE can be useful and effective, it does have a few negative aspects. First, it does not scale well for large sample sizes such as those you would see in some cases of RNA sequencing data. Second, it also does not preserve global data structures in the sense that similarities across different clusters are not well maintained. Another popular model that attempts to address some of these concerns and utilizes a similar approach to t-SNE is known as UMAP. Let's explore this model in the following section.

UMAP

The **UMAP** model is a popular algorithm used for the reduction of dimensions and visualizing of data based on manifold learning techniques, similar to that of t-SNE. There are three main assumptions that the algorithm is founded on, as described on their main website (`https://umap-learn.readthedocs.io/en/latest`) and outlined here:

- The dataset is uniformly distributed on a Riemannian manifold.

- The Riemannian metric is locally constant.

- The manifold is locally connected.

Although UMAP and t-SNE are quite similar, there are a few key differences between them. One of the most important differences relates to the idea of similarity preservation. The UMAP model claims to preserve both local and global data in the sense that both local and global—or inter-cluster and intra-cluster—information is maintained. We can see a graphical representation of this concept here:

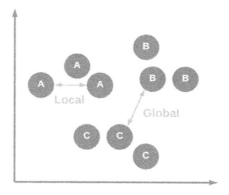

Figure 6.24 – Graphical representation of local and global similarities

Let's now go ahead and apply UMAP on our single-cell RNA dataset. We can begin by importing the umap library and instantiating a new instance of the UMAP model in which we specify the number of components as 2 and the number of neighbors as 5. This second parameter represents the size of a local **neighborhood** used for manifold approximation. The code is illustrated in the following snippet:

```
import umap
data_umap_2d_n5 = umap.UMAP(n_components=2, n_neighbors=5).
fit_transform(dfx)
```

We can then go ahead and plot the data using `seaborn`, as follows:

```
sns.scatterplot(x=data_umap_2d_n5[:,0], y=data_umap_2d_n5[:,1],
hue=dfy.annotation, style=dfy.annotation, markers = ["o", "s",
"v"])
```

Upon executing the code, we yield the following output:

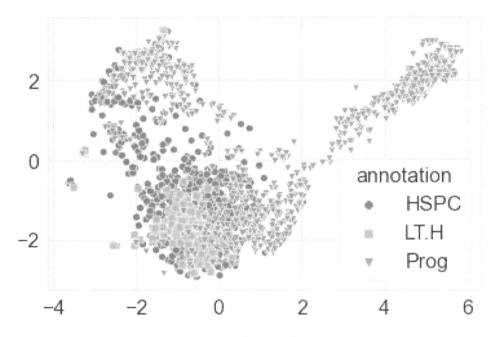

Figure 6.25 – Scatter plot of the UMAP results

Once again, quite the visual! We can see in this depiction relative to t-SNE that some clusters have moved around. If you recall in t-SNE, the majority of the data was pulled together with no regard as to how similar clusters were to one another. Using UMAP, this information is preserved, and we are able to get a better sense of how these clusters relate to one another. Notice the spread of the data relative to its depiction in t-SNE. Similar to t-SNE, we can see that some *groups* of points are clustered together in different neighborhoods.

In summary, UMAP is a powerful model similar to t-SNE in which both local and global information is preserved when it comes to neighborhoods or clusters. Most commonly used for visualizations, this model is an excellent way to gain a sense of the *look* and *feel* of any high-dimensional dataset in just a few lines of code.

Summary

Over the course of this chapter, we gained a strong and high-level understanding of the field of UL, its uses, and its applications. We then explored a few of the most popular ML methods as they relate to clustering and DR. Within the field of clustering, we looked over some of the most commonly used models such as hierarchical clustering, K-Means clustering, and GMMs. We learned about the differences between Euclidean distances and probabilities and how they relate to model predictions. In addition, we also applied these models to the `Wisconsin Breast Cancer` dataset and managed to achieve relatively high accuracy in a few of them. Within the field of DR, we gained a strong understanding of the significance of the field as it relates to the COD. We then implemented a number of models such as PCA, SVD, t-SNE, and UMAP using the *single-cell RNA* dataset in which we managed to reduce more than 1,400 columns down to 2. We then visualized our results using `seaborn` and examined the differences between the models. Over the course of this chapter, we managed to develop our models without the use of labels, which we used only for comparison after the development process.

Over the course of the next chapter, we will explore the field of SML, in which we use data in addition to its labels to develop powerful predictive models.

7
Supervised Machine Learning

As you begin to progress your career and skill set in the field of data science, you will encounter many different types of models that fall into one of the two categories of either supervised or unsupervised learning. Recall that in applications of unsupervised learning, models are generally trained to either cluster or transform data in order to group or reshape data to extract insights when labels are not available for the given dataset. Within this chapter, we will now discuss the applications of **supervised learning** as they apply to the areas of classification and regression to develop powerful predictive models to make educated guesses about a dataset's labels.

Over the course of this chapter, we will discuss the following topics:

- Understanding supervised learning
- Measuring success in supervised machine learning
- Understanding classification in supervised machine learning
- Understanding regression in supervised machine learning

With our objectives in mind, let's now go ahead and get started.

Understanding supervised learning

As you begin to explore data science either on your own or within an organization, you will often be asked the question, *What exactly does supervised machine learning mean?* Let's go ahead and come up with a definition. We can define supervised learning as a general subset of machine learning in which data, like its associated labels, is used to train models that can learn or generalize from the data to make predictions, preferably with a high degree of certainty. Thinking back to *Chapter 5, Introduction to Machine Learning,* we can recall the example we completed concerning the breast cancer dataset in which we classified tumors as being either malignant or benign. This example, alongside the definition we created, is an excellent way to learn and understand the meaning behind supervised learning.

With the definition of supervised machine learning now in our minds, let's go ahead and talk about its different subtypes, namely, classification and regression. If you recall, **classification** within the scope of machine learning is the act of predicting a **category** for a given set of data, such as classifying a tumor as malignant or benign, an email as spam or not spam, or even a protein as alpha or beta. In each of these cases, the model will output a **discrete** value. On the other hand, **regression** is the prediction of an **exact value** using a given set of data, such as the lipophilicity of a small molecule, the isoelectric point of a **Monoclonal Antibody (mAb)**, or the LCAP of an LCMS peak. In each of these cases, the model will output a **continuous** value.

Many different models exist within the two categories of supervised learning. Within the scope of this book, we will focus on four main models for each of these two categories. When it comes to classification, we will discuss **K-Nearest Neighbor (KNN)**, **Support Vector Machines (SVMs)**, **decision trees**, and **random forests**, as well as XGBoost classification. When it comes to regression, we will discuss **linear regression**, **logistic regression**, **random forest regression**, and even **gradient boosting regression**. We can see these depicted in *Figure 7.1*:

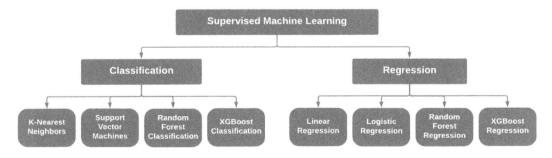

Figure 7.1 – The two areas of supervised machine learning

Our main objective in each of these models is to train a new instance of that model for a particular dataset. We will **fit** our model with the data, and **tune** or adjust the parameters to give us the best outcomes. To determine what the best outcomes should be, we will need to know how to measure success within our models. We will learn about that in the following section.

Measuring success in supervised machine learning

As we begin to train our supervised classifiers and regressors, we will need to implement a few ways to determine which models are performing better, thus allowing us to effectively tune the model's parameters and maximize its performance. The best way to achieve this is to understand what success looks like ahead of time before diving into the model development process. There are many different methods for measuring success depending on the situation. For example, accuracy can be a good metric for classifiers, but not regressors. Similarly, a business case for a classifier may not necessarily require accuracy to be the primary metric of interest. It simply depends on the situation at hand. Let's take a look at some of the most common metrics used for each of the fields of **classification** and **regression**.

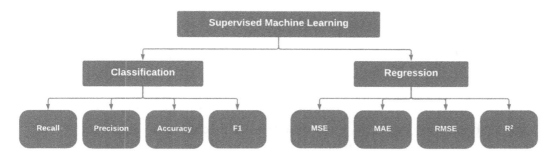

Figure 7.2 – Common success metrics for regression and classification

Although there are many other metrics you can use for a given scenario, the eight listed in *Figure 7.2* are some of the most common ones you will likely encounter. Selecting a given metric can be difficult as it should always align with the given use case. Let's go ahead and explore this when it comes to classification.

Measuring success with classifiers

Take, for example, the tumor dataset we have worked with thus far. We defined our success metric as accuracy, and therefore maximizing accuracy was our model's main training objective. This, however, is not always the case, and the success metric you choose to use will almost always be dependent on both the model and the business problem at hand. Let's go ahead and take a closer look at some metrics commonly used in the data science space and define them:

- **Accuracy**: A measurement that agrees closely with the accepted value

- **Precision**: A measurement that agrees with other measurements in the sense that they are similar to one another

An easier way to think about accuracy and precision is by picturing the results displayed using a bullseye depiction. The difference between precision and accuracy is in the sense of both how close the results are to one another, and how close the results are to their true or actual values, respectively. We can see a visual depiction of this in *Figure 7.3*:

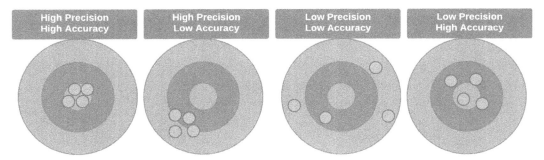

Figure 7.3 – Graphical illustration of the difference between accuracy and precision

In addition to the visual depiction, we can also think of precision as a calculation representing the results as subsets relative to a total population. In this case, we will also need to define a new metric known as **recall**. We can think of recall and precision mathematically in the context of positive and negative results through what is known as a **confusion matrix**. When comparing the results of a prediction relative to the actual results, we can get a good sense of the model's performance by comparing a few of these values. We can see a visual depiction of this in *Figure 7.4*:

		Actual	
		Positive	**Negative**
Predicted	**Positive**	True Positive	False Positive
	Negative	False Negative	True Negative

Figure 7.4 – Graphical illustration of a confusion matrix

With this table in mind, we can define **recall** as the fraction of fraudulent cases that a given model identifies, or, from a mathematical perspective, we can define it as follows:

$$Recall = \frac{true\ positives}{true\ positives + false\ negatives}$$

Whereas we can define **precision** in the same context as follows:

$$Precision = \frac{true\ positives}{true\ positives + false\ positives}$$

We can visualize accuracy, precision, and recall in a similar manner in the following diagram, in which each metric represents a specific calculation of the overall results:

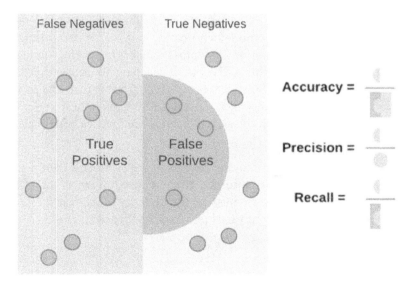

Figure 7.5 – Graphical illustration explaining accuracy, precision, and recall

Finally, there is one last commonly used metric that is usually considered to be a loose *combination* of precision and recall known as the **F1 score**. We can define the *F1 score* as follows:

$$F1\ Score = 2 * \frac{precision * recall}{precision + recall}$$

So how do you determine which metric to use? There is no *best* metric that you should always use as it is highly dependent on each situation. When determining the best metric, you should always ask yourself, *What is the main objective for the model, as well as the business?* In the eyes of the model, accuracy may be the best metric. On the other hand, in the eyes of the business, recall may be the best metric.

Ultimately, recall could be considered more useful when overlooked cases (defined above as false negatives) are more important. Consider, for example, a model that is predicting a patient's diagnosis – we would likely care more about false negatives than false positives. On the other hand, precision can be more important when false positives are more costly to us. It all depends on the given business case and requirements. So far, we have investigated success as it relates to classification, so let's now investigate these ideas as they relate to regression.

Measuring success with regressors

Although we have not yet taken a deep dive into the field of regression, we have defined the main idea as the development of a model whose output is a continuous numerical value. Take, for example, the molecular toxicity dataset containing many columns of data whose values are all continuous floats. Hypothetically, you could use this dataset to make predictions on the **Total Polar Surface Area** (**TPSA**). In this case, the metrics of accuracy, precision, and recall would not be the most useful to us to best understand the performance of our models. Alternatively, we will need some metrics better catered to continuous values.

One of the most common metrics for defining success in many models (not necessarily machine learning) is the **Pearson correlation coefficient**, also known as **R2**. This calculation is a common method used to measure the linearity of data, as it represents the proportion of variance in the dependent variable. We can define **R2** as follows:

$$R^2 = 1 - \frac{\Sigma(y_i - \hat{y})^2}{\Sigma(y_i - \bar{y})^2}$$

In this equation, $\hat{y}$ is the predicted value, and $\bar{y}$ is the mean value.

Take, for example, a dataset in which experimental (actual) values were known, and predicted values were calculated. We could plot the graphs of these values against one another and measure the correlation. In theory, a perfect model would have an ideal correlation (as close to a value of 1.00 as possible). We can see a depiction of high and low correlation in *Figure 7.6*:

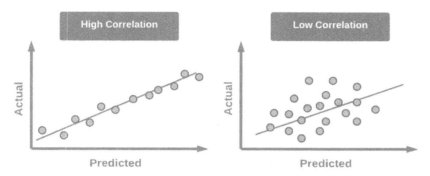

Figure 7.6 – Difference between high and low correlation in scatter plots

Although this metric can give you a good estimate of a model's performance, there are a few others that can give you a better sense of the model's error: **Mean Absolute Error (MAE)**, **Mean Squared Error (MSE)**, and **Root Mean Squared Error (RMSE)**. Let's go ahead and define these:

- **MAE**: The average of the absolute differences between the actual and predicted values in a given dataset. This measure tends to be more robust when handling datasets with outliers:

$$MAE = \frac{1}{N} \sum_{i=1}^{N} |y_i - \hat{y}|$$

In which $\hat{y}$ is the predicted value, and y is the mean value.

- **MSE**: The average of the squared differences between the actual and predicted values in a given dataset:

$$MSE = \frac{1}{N} \sum_{i=1}^{N} (y_i - \hat{y})^2$$

- **RMSE**: Square root of the MSE to measure the standard deviation of the values. This metric is commonly used to compare regression models against each other:

$$RMSE = \sqrt{MSE} = \sqrt{\frac{1}{N}\sum_{i=1}^{N}(y_i - \hat{y})^2}$$

When it comes to regression, there are many different metrics you can use depending on the given situation. In most regression models, RMSE is generally used to compare the performance of multiple models as it is quite simple to calculate and differentiable. On the other hand, datasets with outliers are generally compared to one another using MSE and MAE. Now that we have gained a better sense of measuring success through various metrics, let's now go ahead and explore the area of classification.

Understanding classification in supervised machine learning

Classification models in the context of machine learning are supervised models whose objectives are to classify or categorize items based on previously learned examples. You will encounter classification models in many forms as they tend to be some of the most common models used in the field of data science. There are three main types of classifiers that we can develop based on the outputs of the model.

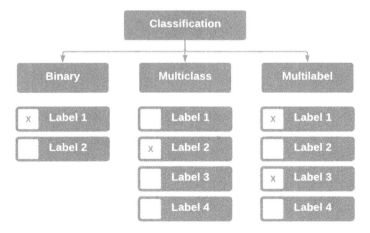

Figure 7.7 – The three types of supervised classification

The first type is known as a **binary classifier**. As the name suggests, this is a classifier that predicts in a binary fashion in the sense that an output is one of two options, such as emails being spam or not spam, or molecules being toxic or not toxic. There is no third option in either of these cases, rendering the model a binary classifier.

The second type of classifier is known as a **multiclass classifier**. This type of classifier is trained on more than two different outputs. For example, many types of proteins can be classified based on structure and function. Some of these examples include structural proteins, enzymes, hormones, storage proteins, and toxins. Developing a model that would predict the type of protein based on some of the protein's characteristics would be regarded as a multiclass classifier in the sense that each row of data could have only one possible class or output.

Finally, we also have **multilabel classifiers**. These classifiers, unlike their multiclass counterparts, are able to predict multiple outputs for a given row of data. For example, when screening patients for clinical trials, you may want to build patient profiles using many different types of labels, such as gender, age, diabetic status, and smoker status. When trying to predict what a certain group of patients might look like, we need to be able to predict all of these labels.

Now that we have broken down classification into a few different types, you are likely thinking about the many different areas in projects you are working on where a classifier may be of great value. The good news here is that many of the standard or popular classification models we are about to explore can be easily recycled and fitted with new data. As we begin to explore the many different models in the following section, think about the projects that you are working on and the datasets you have available, and which models they may fit the best.

Exploring different classification models

As we explore a number of machine learning models, we will test out their performances on a new dataset concerning *single-cell RNA sequences*, published by *Nestorowa et al.* in 2016. We will focus on using this structured dataset in order to develop a number of different classifiers. Let's go ahead and import the data and prepare it for the classification models. First, we will go ahead and import our dataset of interest using the `read_csv()` function in `pandas`:

```
dfx = pd.read_csv("../../datasets/single_cell_rna/nestorowa_
  corrected_log2_transformed_counts.txt", sep=' ',  )
```

Next, we will use the index to isolate our labels (classes) for each of the rows, using the first four characters of each row:

```
dfx['annotation'] = dfx.index.str[:4]
y = dfx["annotation"].values.ravel()
```

We can use the `head()` function to take a look at the data. What we will notice is that there is more than 3,992 columns' worth of data. As any good data scientist knows, developing models with too many columns will lead to many inefficiencies, and therefore it would be best to reduce these down using an unsupervised learning technique, such as **PCA**. Prior to applying PCA, we will need to scale or normalize our dataset using the `StandardScaler` class in `sklearn`:

```
from sklearn.preprocessing import StandardScaler
scaler = StandardScaler()
X_scaled = scaler.fit_transform(dfx.drop(columns =
["annotation"]))
```

Next, we can apply PCA to reduce our dataset from 3,992 columns down to 15:

```
from sklearn.decomposition import PCA
pca = PCA(n_components=15, svd_solver='full')
pca.fit(X_scaled)
data_pca = pca.fit_transform(X_scaled)
```

With the data now in a much more reduced state, we can check the **explained variance ratio** to see how this compares with the original dataset. We will see that the sum of all columns totals 0.17, which is relatively low. We will want to aim for a value around 0.8, so let's go ahead and increase the total number of columns in order to increase the percentage of variance:

```
pca = PCA(n_components=900, svd_solver='full')
pca.fit(X_scaled)
data_pca = pca.fit_transform(X_scaled)
```

With the PCA model applied, we managed to reduce the total number of columns by roughly 77%.

With this completed, we are now prepared to split our dataset using the `train_test_split()` class:

```
from sklearn.model_selection import train_test_split
X_train, X_test, y_train, y_test = train_test_split(data_pca,
y, test_size=0.33)
```

With the dataset now split into training and test sets, we are now ready to begin the classification model development process!

K-Nearest Neighbors

One of the classic, easy-to-develop, and most commonly discussed classification models is known as the **KNN** model, first developed by Evelyn Fix and Joseph Hodges in 1951. The main idea behind this model is determining class membership based on proximity to the closest neighbors. Take, for example, a 2D **binary** dataset in which items are classified as either A or B. As a new dataset is added, the model will determine its membership or class based on its proximity (usually **Euclidean**) to other items in the same dataset.

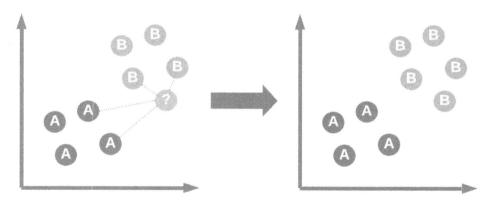

Figure 7.8 – Graphical representation of the KNN model

KNN is regarded as one of the easiest machine learning models to develop and implement given its simple nature and clever design. The model, although simple in application, does require some tuning in order to be fully effective. Let's go ahead and explore the use of this model for the single-cell RNA classification dataset:

1. We can begin by importing the `KNeighborsClassifier` model from `sklearn`:

    ```
    from sklearn.neighbors import KNeighborsClassifier
    ```

2. Next, we can instantiate a new instance of this model in Python, the number of neighbors to a value of 5, and fit the model to our training data:

```
knn = KNeighborsClassifier(n_neighbours=5)
knn.fit(X_train, y_train)
```

3. With the model fit, we can now go ahead and predict the outcomes of the model and set those to a variable we will call y_pred, and finally use the classification_report function to see the results:

```
y_pred = knn.predict(X_test)
print(classification_report(y_test, y_pred))
```

Using the classification report function, we can get a sense of the precision, recall, and F1 scores for each of the three classes. We can see that the precision was relatively high for the LT.H class, but slightly lower for the other two. Alternatively, recall was very low for the LT.H class, but quite high for the Prog class. In total, an average precision of 0.63 was calculated for this model:

```
             precision    recall  f1-score   support

      HSPC       0.66      0.13      0.21       229
      LT.H       1.00      0.04      0.07        54
      Prog       0.52      1.00      0.68       260

avg / total      0.63      0.53      0.42       543
```

Figure 7.9 – Results of the KNN model

With these results in mind, let's go ahead and tune one of the parameters, namely, the n_neighbours parameter in the range of 1-10.

4. We can use a simple for loop to accomplish this:

```
for i in range(1,10):
    knn = KNeighborsClassifier(n_neighbors=i)
    knn.fit(X_train, y_train)
    y_pred = knn.predict(X_test)
    print("n =", i, "acc =", accuracy_score(y_test, y_
pred))
```

If we take a look at the results, we can see the number of neighbors as well as the overall model accuracy. Immediately, we notice that the option value based on this metric alone is n=2, giving an accuracy of approximately 60%.

```
n = 1 acc = 0.5524861878453039
n = 2 acc = 0.6077348066298343
n = 3 acc = 0.5395948434622467
n = 4 acc = 0.585635359116022
n = 5 acc = 0.5340699815837937
n = 6 acc = 0.569060773480663
n = 7 acc = 0.5285451197053407
n = 8 acc = 0.5524861878453039
n = 9 acc = 0.5193370165745856
```

Figure 7.10 – Results of the KNN model at different neighbors

KNN is one of the simplest and fastest models for the development of classifiers; however, it is not always the best model for a complex dataset such as this one. You will notice that the results varied heavily from class to class, indicating the model was not able to effectively distinguish between them based on their proximity to other members alone. Let's go ahead and explore another model known as an SVM, which tries to classify items in a slightly different way.

Support Vector Machines

SVMs are a class of supervised machine learning models commonly used for both classification and regression, first developed in 1992 by AT&T Bell Laboratories. The main idea behind SVMs is the ability to separate classes using a **hyperplane**. There are three main types of SVMs that you will likely hear about in discussions or encounter in the data science field: linear SVMs, polynomial SVMs, and RBF SVMs.

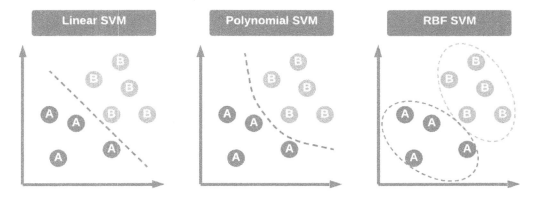

Figure 7.11 – Visual explanation of the different SVMs

The main idea behind the three models lies in how the classes are separated. For example, in **linear** models, the hyperplane is a linear line separating the two classes from each other. Alternatively, the hyperplane may consist of a **polynomial**, allowing the model to account for non-linear features. Finally, and most popularly, the model can use a **Radial Basis Function** (**RBF**) to determine a datapoint's membership, which is based on two parameters, **gamma** and **C**, which account for the decision region, how it is spread out, and the penalty for a misclassification. With this in mind, let's now take a closer look at the idea of a hyperplane.

The **hyperplane** is a function that attempts to clearly define and allow for the differentiation between classes in either a linear or non-linear fashion. The hyperplane can be described mathematically as follows:

$$y = w_0 + w_1x_1 + w_2x_2 + w_3x_3 + \cdots + w_nx_n$$

$$= w_0 + \sum_{i=1}^{m} w_ix_i$$

$$= w_1 + w^TX$$

$$= b + w^TX$$

In which w_i are the vectors, b is the bias term, and w_0 are the variables.

Taking a quick break from the RNA dataset, let's go ahead and demonstrate the use of a linear support vector using the enrollment dataset in Python – a dataset concerning patient enrolment in which respondent data was summarized via **PCA** into two features. The three possible labels within this dataset are `Likely`, `Very Likely`, or `Unlikely` to enroll. The main objective of a **linear SVM** is to *draw a line* clearly separating the data based on class.

Before we begin using SVM, let's go ahead and import our dataset:

```
df = pd.read_csv("../datasets/dataset_enrollment_sd.csv")
```

For simplicity, let's eliminate the `Likely` class and keep the `Very Likely` and `Unlikely` classes:

```
dftmp = df[(df["enrollment_cat"] != "Likely")]
```

Let's go ahead and draw a line separating the data shown in the scatter plot:

```
plt.figure(figsize=(15, 6))
xfit = np.linspace(-90, 130)
    sns.scatterplot(dftmp["Feature1"],
                    dftmp["Feature2"],
                    hue=dftmp["enrollment_cat"].values,
                    s=50)
    for m, b in [(1, -45),]:
        plt.plot(xfit, m * xfit + b, '-k')
    plt.xlim(-120, 150);
    plt.ylim(-100, 60);
```

Upon executing this code, this yields the following diagram:

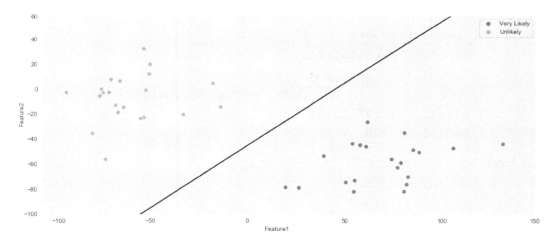

Figure 7.12 – Two clusters separated by an initial SVM hyperplane

Notice within the plot that this linear line could be drawn in multiple ways with different slopes, yet still successfully separate the two classes within the dataset. However, as new datapoints begin to encroach toward the middle ground between the two clusters, the slope and location of the hyperplane will begin to grow in importance. One way to address this issue is by defining the slope and location of the plane based on the closest datapoints. If the line contained a margin of width x in relation to the closest datapoints, then a more improved **hyperplane** could be constructed. We can construct this using the `fill_between` function, as portrayed in the following code:

```
plt.figure(figsize=(15, 6))
xfit = np.linspace(-110, 180)
sns.scatterplot(dftmp["Feature1"],
                dftmp["Feature2"],
                hue=dftmp["enrollment_cat"].values,
                s=50)
for m, b, d in [(1, -45, 60),]:
yfit = m * xfit + b
plt.plot(xfit, yfit, '-k')
    plt.fill_between(xfit, yfit - d,
                yfit + d, edgecolor='none',
                    color='#AAAAAA', alpha=0.4)
plt.xlim(-120, 150);
plt.ylim(-100, 60);
```

Upon executing this code, this yields the following figure:

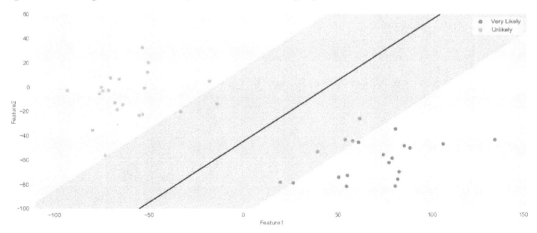

Figure 7.13 – Two clusters separated by an initial SVM hyperplane with specified margins

The datapoints from both classes that are within the margin width of the hyperplane are known as **support vectors**. The main intuition is the idea that the further away the support vectors are from the hyperplane, the higher the probability that a correct class is identified for a new datapoint.

We can train a new SVM classifier using the SVC class from the `scikit-learn` library. We begin by importing the class, splitting the data, and then training a model using the dataset:

```
from sklearn.model_selection import train_test_split
X_train, X_test, y_train, y_test =
                   train_test_
split(dftmp[["Feature1","Feature2"]],
                              dftmp["enrollment_cat"].
values,
                              test_size = 0.25)
from sklearn.svm import SVC
model = SVC (kernel='linear', C=1E10, random_state = 42)
model.fit(X_train, y_train)
```

With that, we have now fitted our model to the dataset. As a final step, we can show the scatter plot, identify the hyperplane, and also specify which datapoints were the support vectors for this particular example:

```
plt.figure(figsize=(15, 6))
sns.scatterplot(dftmp["Feature1"],
               dftmp["Feature2"],
               hue=dftmp["enrollment_cat"].values, s=50)
plot_svc_decision_function(model);
for j, k in model.support_vectors_:
     plt.plot([j], [k], lw=0, ='o', color='red',
            markeredgewidth=2, markersize=20,
            fillstyle='none')
```

Upon executing this code, this yields the following figure:

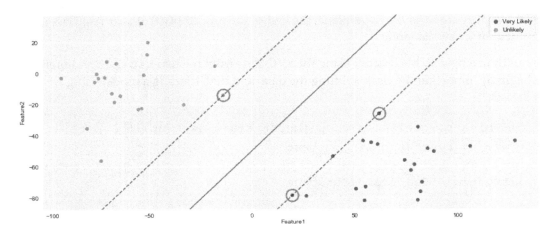

Figure 7.14 – Two clusters separated by an initial SVM hyperplane with select support vectors

Now that we have gained a better understanding of how SVMs operate in relation to their hyperplanes using this basic example, let's go ahead and test out this model using the single-cell RNA classification dataset we have been working with.

Following roughly the same steps as the KNN model, we will now implement the SVM model by first importing the library, instantiating the model with a linear kernel, fitting our training data, and subsequently making predictions on the test data:

```python
from sklearn.svm import SVC
svc = SVC(kernel="linear")
svc.fit(X_train, y_train)
y_pred = svc.predict(X_test)
print(classification_report(y_test, y_pred))
```

Upon printing the report, this yields the following results:

```
               precision    recall  f1-score   support

        HSPC        0.83      0.90      0.87       229
        LT.H        0.82      0.67      0.73        54
        Prog        0.93      0.90      0.92       260

   avg / total      0.88      0.88      0.88       543
```

Figure 7.15 – Results of the SVM model

We can see that the model was, in fact, quite robust, with our dataset yielding some high metrics and giving us a total average precision of 88%. SVMs are fantastic models to use with complex datasets as their main objective is to separate data via a hyperplane. Let's now explore a model that takes a very different approach by using decision trees to arrive at final results.

Decision trees and random forests

Decision trees are one of the most popular and commonly used machine learning models when it comes to structured datasets for both classification and regression. Decision trees consist of three elements: **nodes**, **edges**, and **leaf nodes**.

Nodes generally consist of a question allowing for the process to split into an arbitrary number of child nodes, shown in orange in the following diagram. The root node is the first node that the entire tree is referenced through. Edges are the connections between nodes shown in blue. When nodes have no children, then this final destination is called a leaf, shown in green. In some cases, a decision tree will have nodes containing the same parent – these are called sibling nodes. The more nodes there are in a tree, the *deeper* the tree is said to be. The depth of the decision tree is a measure of *complexity*.

A tree that is not complex enough will not arrive at an accurate result, and a tree that is too complex will be overtrained. Identifying a good balance is one of the primary objectives in the training process. Using these elements, you can construct a decision tree, allowing for processes to flow from the top to the bottom, thereby arriving at a particular destination or *decision*:

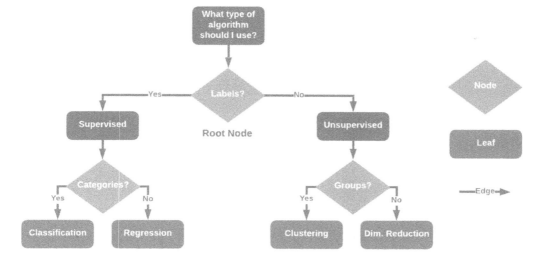

Figure 7.16 – Illustration of decision trees when it comes to nodes, leaves, and edges

Decision trees operate in quite a genius way. We begin with our initial dataset in which all datapoints are labeled and ready to go. The first objective is to split the dataset using a decision boundary that is the most informative – which, in this case, is at $y=m$. This has successfully isolated a class from the two others; however, the two others are still not yet isolated from one another. The algorithm then splits the dataset again at $x = n$, thus completely separating the three clusters. If more clusters were present, this process would iteratively and recursively continue until all classes are optimally separated. We can see a visual representation of this in *Figure 7.17*:

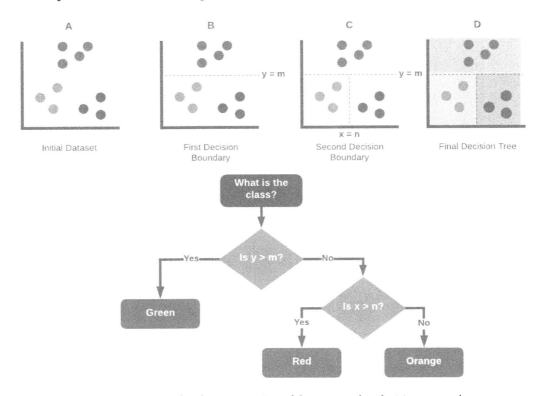

Figure 7.17 – A graphical representation of the process that decision trees take

Decision trees determine where and how to split the data using various splitting criteria, known as attribute selection measures. These prominent attribute selection measures include:

- Information gain
- Gain ratio
- Gini index

Let's now take a closer look at these three items.

Information gain is an attribute concerning the amount of information required to further describe the tree. This attribute minimizes the information needed for data classification while utilizing the least randomness in the partitions. Think of information gain of a random variable determined from the observation of a random variable, A, as a function such that:

$$IG_{XA}(X, a) = D_{KL}(P_X(x|a) \parallel P_X(x|I))$$

Broadly speaking, the information gain is the change in entropy (information entropy) H such that:

$$IG(T, a) = H(T) - H(T|a)$$

In which $H(T|a)$ represents the conditional entropy of T given the attribute a. In summary, information gain answers the question, *How much information do we obtain from this variable given another variable?*

On the other hand, the **gain ratio** is the information gain relative to the intrinsic information. In other words, this measure is biased toward tests that result in many outcomes, thus forcing a preference in favor of features of this nature. The gain ratio can be represented as:

$$GainRatio(A) = \frac{Gain(A)}{SplitInformation(A)}$$

In which $SplitInformation$ is represented as:

$$SplitInformation_A(D) = -\sum_{j=1}^{v} \frac{|D_j|}{|D|} * \log_2 \left(\frac{|D_j|}{|D|} \right)$$

In summary, the gain ratio penalizes variables with more distinct values, which will help decide the next split at the next level.

Finally, we arrive at the **Gini index**, which is an attribute selection measure representing how often a randomly selected element is incorrectly labeled. The Gini index can be calculated by subtracting the sum of square probabilities of each class:

$$Gini(D) = 1 - \sum_{i=1}^{m} p_i^2$$

This methodology in determining a split naturally favors larger partitions as opposed to information gain, which favors smaller ones. The objective of any data scientist is to explore different methods with your dataset and determine the best path forward.

Now that we have a much more detailed explanation of decision trees and how the model operates, let's now go ahead and implement this model using the previous single-cell RNA classification dataset:

```
from sklearn.tree import DecisionTreeClassifier
dtc = DecisionTreeClassifier(max_depth=4)
dtc.fit(X_train, y_train)
y_pred = dtc.predict(X_test)
print(classification_report(y_test, y_pred))
```

Upon printing the report, this yields the following results:

```
              precision    recall  f1-score   support

       HSPC        0.71      0.82      0.76       229
       LT.H        0.57      0.30      0.39        54
       Prog        0.86      0.83      0.84       260

  avg / total      0.77      0.77      0.76       543
```

Figure 7.18 – Results of the decision tree classifier

We can see that the model, without any tuning, was able to deliver a total precision score of 77% using a `max_depth` value of 4. Using the same method as the *KNN* model, we can iterate over a range of `max_depth` values to determine the optimal value. Doing so would result in an ideal `max_depth` value of 3, yielding a total precision of 82%.

As we begin to train many of our models, one of the most common issues that we will face is **overfitting** our data in one way or another. Take, for example, a decision tree model that was very finely tuned for a specific selection of data since decision trees are built on an entire dataset using the features and variables of interest. In this case, the model may be prone to overfitting. On the other hand, another model known as a **random forest** is built using many different decision trees to help mitigate any potential overfitting.

Random forests are robust ensemble models based on decision trees. Where decision trees are generally designed to develop a model using the dataset as a whole, random forests randomly select features and rows and subsequently build multiple decision trees that are then averaged in their weights. Random forests are powerful models given their ability to limit overfitting while avoiding a substantial increase in error due to **bias**. We can see a visual representation of this in *Figure 7.19*:

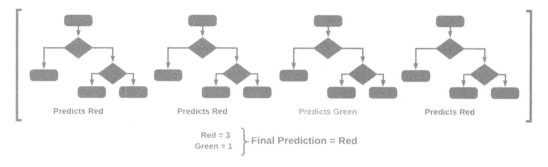

Figure 7.19 – Graphical explanation of random forest models

There are two main ways in which random forests can help reduce **variance**. The first method is by training on different samples of data. Consider the preceding example using the patient enrollment data. If the model was trained on samples not containing those *in between* the clusters, then determining the score on the test set will result in significantly lower accuracy.

The second method involves using a random subset of features to train on, allowing for the determination of the concept of feature importance within the model. Let's go ahead and take a look at this model using the single-cell RNA classification dataset:

```
from sklearn.ensemble import RandomForestClassifier
clf = RandomForestClassifier(n_estimators=1000)
clf.fit(X_train, y_train)
y_pred = clf.predict(X_test)
print(classification_report(y_test, y_pred))
```

Upon printing the report, this yields the following results:

```
              precision    recall  f1-score   support

        HSPC       0.71      0.94      0.81       229
        LT.H       0.00      0.00      0.00        54
        Prog       0.93      0.85      0.89       260

   avg / total       0.74      0.81      0.77       543
```

Figure 7.20 – Results of the random forest model

We can immediately observe that the model has a precision of roughly 74%, slightly lower than the decision tree above, indicating that the tree may have overfitted the data slightly.

Random forest models are very commonly used in the biotechnology and life sciences industries given their remarkable methods of avoiding overfitting, as well as their ability to develop predictive models with smaller datasets. Many applications of machine learning within the biotech space generally suffer from a concept known as the **low-N** problem, in the sense that use cases exist, but little to no data has been collected or organized with which to develop a model. Random forests are commonly used for applications in this space given their ensemble nature. Let's now take a look at a very different model that splits data not based on decisions, but based on statistical probability instead.

Extreme Gradient Boosting (XGBoost)

Over the last few years, a number of robust machine learning models have begun to enter the data science space, thus changing the machine learning landscape quite effectively – one of these models being the **Extreme Gradient Boosting** (**XGBoost**) model. The main idea behind this model is that it is an implementation of **Gradient Boosted Models** (**GBMs**), specifically, *decision trees*, in which speed and performance were highly optimized. Because of the highly efficient and highly effective nature of this model, it began to dominate many areas of data science and eventually became the go-to algorithm for many data science competitions on Kaggle.com.

There are many reasons why GBMs are so effective with structured/tabular datasets. Let's go ahead and explore three of these reasons:

- **Parallelization**: The *XGBoost* model implements a method known as parallelization. The main idea here is that it can parallelize processes in the construction of each of the trees. In essence, each of the branches of a single tree is trained separately.

- **Tree-Pruning**: Pruning is the process of removing parts of a decision tree deemed to be redundant or not useful to a model. The main idea behind *GBM* pruning, simply stated, is that the GBM model would stop splitting a given node when a negative loss in the split is encountered. *XGBoost*, on the other hand, splits up to the max_depth parameter, and then begins the pruning process backward and eventually removes splits after which there is no longer a positive gain.

- **Regularization**: In the context of tree-based methods overall, regularization is an algorithmic method to define a minimum gain in order to prompt another split in the tree. In essence, regularization shrinks scores, thereby prompting the final prediction to be more conservative, which, in return, helps prevent overfitting within the model.

Now that we have gained a much better understanding of XGBoost and some of the reasons behind its robust performance, let's go ahead and implement this model on our RNA dataset. We will begin by installing the model's library using `pip`:

```
pip install xgboost
```

With the model installed, let's now go ahead and import the model and then create a new instance of the model in which we specify the n_estimators parameter to have a value of 10000:

```
from xgboost import XGBClassifier
xgb = XGBClassifier(n_estimators=10000)
```

Similar to the previous models, we can now go ahead and fit our model with the training datasets and print the results of our predictions:

```
xgb.fit(X_train, y_train)
y_pred = xgb.predict(X_test)
print(classification_report(y_test, y_pred))
```

Upon printing the report, this yields the following results:

```
              precision    recall  f1-score   support

        HSPC       0.81      0.90      0.85       229
        LT.H       0.76      0.54      0.63        54
        Prog       0.92      0.89      0.91       260

 avg / total       0.86      0.86      0.86       543
```

Figure 7.21 – Results of the XGBoost model

With that, we can see that we managed to achieve a precision of 0.86, much higher than some of the other models we tested. The highly optimized nature of this model allows it to be very fast and robust relative to most others.

Over the course of this section, we managed to cover quite a wide scope of **classification** models. We began with the simple **KNN** model, which attempts to predict the class of a new value relative to its closest neighbors. Next, we covered **SVM** models, which attempt to assign labels based on specified boundaries drawn by support vectors. We then covered both **decision trees** and **random forests**, which operate based on nodes, leaves, and splits, and then finally saw a working example of **XGBoost**, a highly optimized model that implements many of the features we saw in other models.

As you begin to dive into the many different models for new datasets, you will likely investigate the idea of automating the model selection process. If you think about it, each of the steps we have taken above could be automated in one way or another to identify which model operates the best under a specific set of metric requirements. Luckily for us, a library already exists that can assist us in this space. Over the course of the following tutorial, we will investigate the use of these models alongside some automated machine learning capabilities on **Google Cloud Platform** (**GCP**).

Tutorial: Classification of proteins using GCP

During this tutorial, we will investigate a number of classification models, followed by an implementation of some automated machine learning capabilities. Our main objective will be to automatically develop a model for the classification of proteins using a dataset from the **Research Collaboratory for Structural Bioinformatics** (**RCSB**) **Protein Data Bank** (**PDB**). If you recall, we used data from RCSB PDB in a previous chapter to plot a 3D protein structure. The dataset we will be working with in this chapter consists of two parts—a **structured dataset** with rows columns, one column of which is the designated classification of each of the proteins, and a series of RNA sequences for each of the proteins. We will save this second set of sequence-based data for analysis in a later chapter and focus on the structured dataset for now.

In this chapter, we will use this structured dataset containing many different types of proteins in an attempt to develop a classifier. Given the large nature of this dataset, we will take this opportunity to move our development environment from our local installation of **Jupyter Notebook** to an online notebook in **GCP**. Before we can do so, we will need to create a new GCP account. Let's go ahead and get started.

Getting started in GCP

Getting started in **GCP** is quite simple, and can be accomplished in a few simple steps:

1. We can begin by navigating to `https://cloud.google.com/` and registering for a new account. You will need to provide a few details, such as your name, email, and a few other items.

2. Once registered, you will be able to navigate to the console by clicking the **Console** button on the upper right-hand side of the page:

Figure 7.22 – A screenshot of the Console button

3. Within the console page, you will be able to see all items relating to your current project, such as general information, resources used, API usage, and even billing:

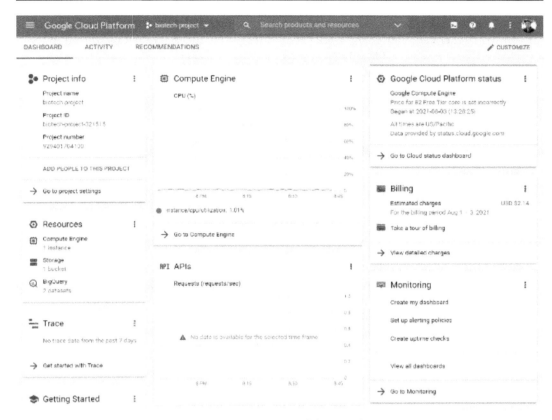

Figure 7.23 – An example of the console page

4. You will likely not have any projects set up yet. In order to create a new project, navigate to the drop-down menu on the upper left-hand side and select the **New Project** option. Give your project a name and then click **CREATE**. You can navigate between different projects using that same drop-down menu:

Figure 7.24 – A screenshot of the project name and location pane in GCP

With that last step completed, you are now all set up to take full advantage of the GCP platform. We will cover a few of the GCP capabilities in this tutorial to get us started in the data science space; however, I highly encourage new users to explore and learn the many tools and resources available here.

Uploading data to GCP BigQuery

There are many different ways in which you can upload data within GCP; however, we will focus on one particular capability unique to the GCP, known as **BigQuery**. The main idea behind BigQuery is that it is a serverless data warehouse with built-in machine learning capabilities that supports the use of queries with the **SQL** language. If you recall, we previously developed and deployed an **AWS RDS** to manage our data using an **EC2** instance as a server, whereas BigQuery, on the other hand, operates using a serverless architecture. We can set up BigQuery and start uploading our data in a few simple steps:

1. Using the navigation menu on the left-hand side of the page, scroll down to the **Products** section, hover over the **BigQuery** option, and select **SQL workspace**. Given that this is the first time you are using this tool, you may need to activate the API. This will be true for all tools that you have never used before:

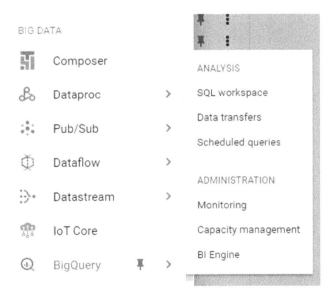

Figure 7.25 – A screenshot of the BigQuery menu in GCP

2. Within this list, you will find the project that you created in the previous section. Click on the options button to the right and select **Create dataset**. In this menu, give the dataset a name, such as `protein_structure_sequence`, leaving all the other options as their default values. You can then click **Create dataset**.

3. On the left-hand menu, you will see the newly created dataset listed under the project name. If you click **Options** followed by **Open**, you will be directed to the dataset's main page. Within this page, you will find information relating to that particular dataset. Let's now go ahead and create a new table here by clicking the **Create Table** option at the top. Change the source to reflect the upload option and navigate to the CSV file pertaining to the protein classifications from RCSB PDB. Give the table a new name, and while leaving all other options as their default values, click **Create Table**:

Create table

Source

Create table from:	Select file:		File format:
Upload ▾	dataset_pdb_no_dups.csv	Browse	CSV ▾

Destination

◉ Search for a project   ○ Enter a project name

Project name	Dataset name	Table type
biotech-project ▾	protein_structure_sequence ▾	Native table ▾

Table name

dataset_pdb_no_dups

Figure 7.26 – A screenshot of the Create table pane in GCP

4. If you navigate back to Explorer, you will see the newly created table listed under the dataset, which is listed under your project:

Figure 7.27 – An example of the table created within a dataset

If you managed to follow all of these steps correctly, you should now have data available for you to use in BigQuery. In the following section, we will prepare a new notebook and start parsing some of our data in this dataset.

Creating a notebook in GCP

In this section, we will create a notebook equivalent to that of the Jupyter notebooks we have been using to carry out our data science work on. We can get a new notebook set up in a few simple steps:

1. In the navigation pane on the left-hand side of the screen, scroll down to the **ARTIFICIAL INTELLIGENCE** section, hover over **AI Platform**, and select the **Notebooks** option. Remember that you may need to activate this API once again if you have not done so already:

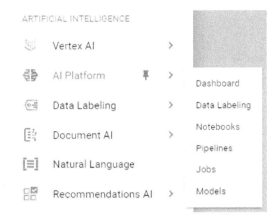

Figure 7.28 – A visual of the AI Platform menu

2. Next, navigate to the top of the screen and select the **New Instance** option. There are many different options available for you depending on your needs. For the purposes of this tutorial, we can select the first option for **Python 3**:

Figure 7.29 – A screenshot of the instance options

If you are familiar with notebook instances and are comfortable customizing them, I recommend creating a customized instance to suit your exact needs.

3. Once the notebook is created and the instance is online, you will be able to see it in the main **Notebook Instances** section. Go ahead and click on the **OPEN JUPYTERLAB** button. A new window will open up containing Jupyter Lab:

Figure 7.30 – A screenshot of the instance menu

4. In the home directory, create a new directory called `biotech-machine-learning` for us to save our notebooks in. Open the directory and create a new notebook by clicking the **Python 3** notebook option on the right:

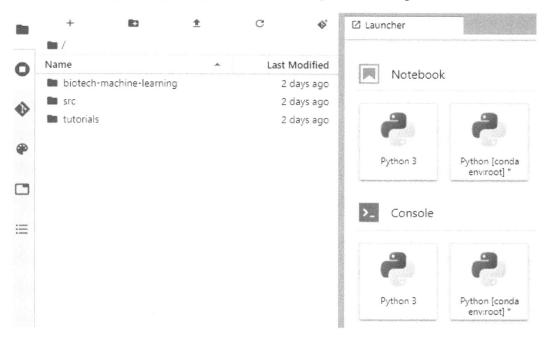

Figure 7.31 – A screenshot of Jupyter Lab on GCP

With the instance now provisioned and the notebook created, you are now all set to run all of your data science models on GCP. Let's now take a closer look at the data and begin training a few machine learning models.

Using auto-sklearn in GCP Notebooks

If you open your newly created notebook, you see the very familiar environment of Jupyter Lab that we have been working with all along. The two main benefits here are that we now have the ability to manage our datasets within this same environment and can provision larger resources to process our data relative to the few **CPUs** and **GPUs** we have on our local machines.

Recall that our main objective for getting to this state is to be able to develop a classification model to correctly classify proteins based on some input features. The dataset we are working with is known as a `real-world` dataset in the sense that it is not well organized, has missing values, may contain too much data, and will need some preprocessing prior to the development of any model.

Let's go ahead and start by importing a few necessary libraries:

```
import pandas as pd
import numpy as np
from google.cloud import bigquery
import missingno as msno
from sklearn.metrics import classification_report
import ast
import autosklearn.classification
```

Next, let's now import our dataset from BigQuery. We can do that directly here in the notebook by instantiating a client using the BigQuery class of the Google Cloud library:

```
client = bigquery.Client(location="US")
print("Client creating using default project: {}".
format(client.project))
```

Next, we can go ahead and query our data using a `SELECT` command in the **SQL** language. We can simply begin by querying all the data in our dataset. In the following code snippet, we will query the data using SQL, and convert the results to a dataframe:

```
query = """
    SELECT *
    FROM `biotech-project-321515.protein_structure_sequence.
dataset_pdb_no_dups`
"""
query_job = client.query(
    query,
```

```
    location="US",
)
df = query_job.to_dataframe()
print(df.shape)
```

Once converted to a more manageable dataframe, we can see that the dataset we are working with is quite extensive, with nearly 140,000 rows and 14 columns of data. Immediately, we notice that one of the columns is called `classification`. Let's take a look at the unique number of classes in this dataset using the `n_unique()` function:

```
df.classification.nunique()
```

We notice that there are 5,050 different classes! That is a lot for a dataset this size, indicating that we may need to reduce this quite heavily before any analysis. Before proceeding any further, let's go ahead and drop any and all potential duplicates:

```
dfx = df.drop_duplicates(["structureId"])
```

Let's now take a closer look at the top 10 classes in our dataset by count:

```
dfx.classification.value_counts()[:10].sort_values().plot(kind
= 'barh')
```

The following figure is yielded from the code, showing us the top 10 classes from this dataset:

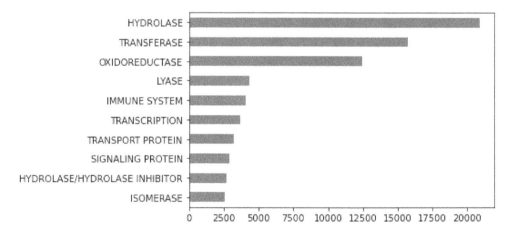

Figure 7.32 – The top 10 most frequent labels in the dataset

Immediately, we notice that there are two or three classes or proteins that account for the vast majority of this data: hydrolase, transferase, and oxidoreductase. This will be problematic for two reasons:

- Data should always be **balanced** in the sense that each of the classes should have a roughly equal number of rows.

- As a general rule of thumb, the ratio of classes to observations should be around 50:1, meaning that with 5,050 classes, we would require around 252,500 observations, which we do not currently have.

Given these two constraints, we can account for both by simply focusing on developing a model using the first three classes. For now, we notice that there are quite a few features available to us regardless of the classes at hand. We can go ahead and take a closer look at the completeness of our features of interest using the msno library:

```
dfx = dfx[["classification", "residueCount", "resolution",
"resolution", "crystallizationTempK", "densityMatthews",
"densityPercentSol", "phValue"]]
msno.matrix(dfx)
```

The following screenshot, representing the completeness of the dataset, is then generated. Notice that a substantial number of rows for the crystallizationTempK feature are missing:

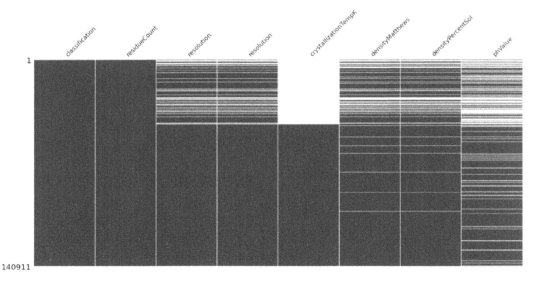

Figure 7.33 – A graphical representation showing the completeness of the dataset

So far within this dataset, we have noted the fact that we will need to reduce the number of classes to the top two classes to avoid an imbalanced dataset, and we will also need to address the many rows of data we are missing. Let's go ahead and prepare our dataset for the development of a few classification models based on our observations. First, we can go ahead and reduce the dataset using a simple `groupby` function:

```
df2 = dfx.groupby("classification").filter(lambda x: len(x) >
14000)
df2.classification.value_counts()
```

If we run a quick check on the dataframe using the `value_counts()` function, we notice that we were able to reduce it down to the top two labels.

Alternatively, we can run this same command in **SQL** using a few clever joins. We can begin with our inner query in which we SELECT the classification and COUNT the `residueCount` feature and the GROUP BY classification. Next, we INNER JOIN that query with the original table setting, classification against classification, but filtering using our WHERE clause:

```
query = """
    SELECT DISTINCT
        dups.*
    FROM (
        SELECT classification, count(residueCount) AS
classCount
        FROM `biotech-project-321515.protein_structure_
sequence.dataset_pdb_no_dups`
        GROUP BY classification
    ) AS sub
    INNER JOIN `biotech-project-321515.protein_structure_
sequence.dataset_pdb_no_dups` AS dups
        ON sub.classification = dups.classification
    WHERE sub.classCount > 14000
"""
query_job = client.query(
    query,
    location="US",
)

df2 = query_job.to_dataframe()
```

Next, we can go ahead and remove the rows of data with missing values using the dropna() function:

```
df2 = df2.dropna()
df2.shape
```

Immediately, we observe that the size of the dataset has been reduced down to 24,179 observations. This will be a sufficient dataset to work with when developing our models. In order to avoid having to process it again, we can write the contents of the dataframe to a new table in the same BigQuery dataset:

```
import pandas_gbq
pandas_gbq.to_gbq(df2, 'protein_structure_sequence.dataset_
pdb_no_dups_cleaned', project_id ='biotech-project-321515',
if_exists='replace')
```

With the data now prepared, let's go ahead and develop a model. We can go ahead and split the input and output data, scale the data using the StandardScaler class, and split the data into test and training sets:

```
X = df2.drop(columns=["classification"])
y = df2.classification.values.ravel()
from sklearn.preprocessing import StandardScaler
scaler = StandardScaler()
X_scaled = scaler.fit_transform(X)
from sklearn.model_selection import train_test_split
X_train, X_test, y_train, y_test = train_test_split(X_scaled,
y, test_size=0.25)
```

For the automation section, we will use a library called autosklearn, which can be installed using the command line via pip:

pip install autosklearn

With the library installed, we can go ahead and import the library and instantiate a new instance of that model. We will then set a few parameters relating to the time we wish to dedicate to this process and give the model a temporary directory to operate in:

```
import autosklearn.classification
automl = autosklearn.classification.AutoSklearnClassifier(
    time_left_for_this_task=120,
    per_run_time_limit=30,
```

```
     tmp_folder='/tmp/autosklearn_protein_tmp5',
)
```

Finally, we can go ahead and fit the model on our data. This process will take a few minutes to run:

```
automl.fit(X_train, y_train, dataset_name='dataset_pdb_no_
dups')
```

When the model is complete, we can take a look at the results by printing the leader board:

```
print(automl.leaderboard())
```

Upon printing the leaderboard, we retrieve the following results:

```
          rank  ensemble_weight               type      cost   duration
model_id
2            1             0.32      random_forest  0.261821  16.188326
3            2             0.08        extra_trees  0.275188  15.251657
10           3             0.22  gradient_boosting  0.288053   9.499943
9            4             0.16      random_forest  0.294737  13.219077
6            5             0.08                mlp  0.397995   3.221838
8            6             0.06          libsvm_svc  0.420217  29.656829
5            7             0.08                mlp  0.421554   9.592515
```

Figure 7.34 – Results of the auto-sklearn model

We can also take a look at the top-performing random_forest model using the get_
models_with_weights() function:

```
automl.get_models_with_weights()[0]
```

We can also go ahead and get a few more metrics by making some predictions using the model and the classification_report() function:

```
predictions = automl.predict(X_test)
print("classification_report:", classification_report(y_test,
predictions))
```

Upon printing the report, this yields the following results:

```
classification_report:               precision    recall  f1-score   support

       HYDROLASE      0.75      0.78      0.76      3434
     TRANSFERASE      0.69      0.66      0.67      2611

        accuracy                         0.73      6045
       macro avg      0.72      0.72      0.72      6045
    weighted avg      0.73      0.73      0.73      6045
```

Figure 7. 35 – Results of the top-performing auto-sklearn model

With that, we managed to develop a machine learning model successfully and automatically for our dataset. However, the model has not yet been fined-tuned or optimized for this task. One challenge that I recommend you complete is the process of tuning the various parameters in this model in an attempt to increase our metrics. In addition, another challenge would be to try and explore some of the other models we learned about and see whether any of them can beat `autosklearn`. Hint: **XGBoost** has always been a great model for structured datasets.

Using the AutoML application in GCP

In the previous section, we used an open source library called `auto-sklearn` that automates the process in which models can be selected using the `sklearn` library. However, as we have seen with the `XGBoost` library, there are many other models out there outside of the `sklearn` API. GCP offers a robust tool, similar to that of `auto-sklearn`, that iterates over a large selection of models and methods to find the most optimal model for a given dataset. Let's go ahead and give this a try:

1. In the navigation menu in GCP, scroll down to the **ARTIFICIAL INTELLIGENCE** section, hover over **Tables**, and select **Datasets**:

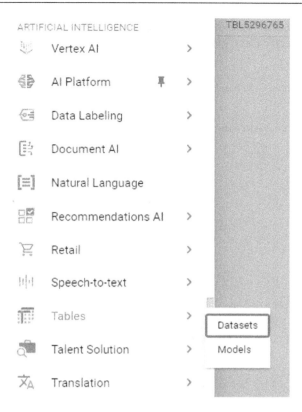

Figure 7.36 – Selecting Datasets from the ARTIFICIAL INTELLIGENCE menu

2. At the top of the page, select the **New Dataset** option. At the time this book was written, the beta implementation of the model was available. Some of the steps will likely have changed in future implementations:

Datasets BETA   ➕ NEW DATASET

Figure 7.37 – A screenshot of the button to create a new dataset

3. Go ahead and give the dataset a name and region and click **Create Dataset**.

4. We have the option to import our dataset of interest either as a raw CSV file or using BigQuery. Go ahead and import our cleaned version of the proteins dataset by specifying `projectID`, `datasetID`, and the table name, and then click **Import**.

5. In the **TRAIN** section, you will have the ability to see the tables within this dataset. Go ahead and specify the **Classification** column as the target column and then click **TRAIN MODEL**:

Figure 7.38 – An example of the training menu

6. The model selection process will take some time to complete. Upon completion, you will be able to see the results of the model under the **Evaluate** tab. Here you will get a sense of the classification metrics we have been working with, as well as a few others.

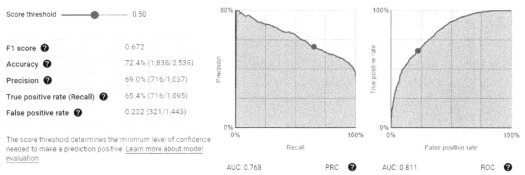

Figure 7.39 – Results of the trained model showing the metrics

GCP AutoML is a powerful tool that you can use to your advantage when handling a difficult dataset. You will find that the implementation of the model is quite robust and generally comprehensive relative to the many options that we, as data scientists, can explore. One of the downsides of **AutoML** is the fact that the final model is not shared with the user; however, the user does have the ability to test new data and use the model later on. We will explore another option similar to **AutoML** in the following section known as **AutoPilot** in **AWS**. Now that we have explored quite a few different models and methods relating to classification, let's go and explore their respective counterparts when it comes to regression.

Understanding regression in supervised machine learning

Regressions are models generally used to determine the relationship or **correlation** between dependent and independent variables. Within the context of machine learning, we define regressions as supervised machine learning models that allow for the identification of correlations between two or more variables in order to **generalize** or learn from historical data to make predictions on new observations.

Within the confines of the **biotechnology** space, we use regression models to predict values in many different areas.

- Predicting the LCAP of a compound ahead of time
- Predicting titer results further upstream
- Predicting the isoelectric point of a monoclonal antibody
- Predicting the decomposition percentages of compounds

Correlations are generally established between two columns. Two columns within a dataset are said to have a strong **correlation** when a dependence is observed. The specific relationship can be better understood using a linear regression model such that:

$$y_i = \beta x_i + a + \varepsilon_i$$

In which y_i is the first feature, x_i is the second feature, ε_i is a small error term, with a and β as constants. Using this simple equation, we can understand our data more effectively, and calculate any correlation. For example, recall earlier we observed a correlation in the toxicity dataset, specifically between the `MolWt` and `HeavyAtoms` features.

The main idea behind any given regression model, unlike its classification counterparts, is to output a continuous value rather than a class or category. For example, we could use a number of columns in the toxicity dataset in an attempt to predict other columns in the same dataset.

There are many different types of regression models commonly used in the data science space. There are linear regressions that focus on **optimizing** a linear relationship between a set of variables, logistic regression that acts more as a binary classifier rather than regressors, and ensemble models that combine the predictive power of several base estimators, among many others. We can see some examples in *Figure 7.40*:

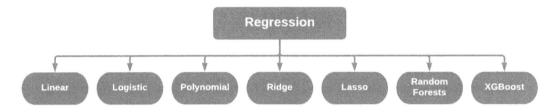

Figure 7.40 – Different types of regression models

Over the course of this section, we will explore a few of these models as we investigate the application of a few regression models using the toxicity dataset. Let's go ahead and prepare our data.

We can begin by importing our libraries of interest:

```
import pandas as pd
import numpy as np

from sklearn.metrics import r2_score
from sklearn.metrics import mean_squared_error

import matplotlib.pyplot as plt
import seaborn as sns
sns.set_theme(color_codes=True)
```

Next, we can go ahead and import our dataset and drop the missing rows. For practice, I suggest you upload this dataset to **BigQuery**, just as we did in the previous section, and query your data using **SQL** via a `SELECT` statement:

```
df = pd.read_csv("../../datasets/dataset_toxicity_sd.csv")
df = df.dropna()
```

Next, we can split our data into input and output values, and scale our data using the `MinMaxScaler()` class from `sklearn`:

```
X = df[["Heteroatoms", "MolWt", "HeavyAtoms", "NHOH",
"HAcceptors", "HDonors"]]
y = df.TPSA.values.ravel()
from sklearn.preprocessing import MinMaxScaler
scaler = MinMaxScaler()
X_scaled = scaler.fit_transform(X)
```

Finally, we can split the dataset into training and test data:

```
from sklearn.model_selection import train_test_split
X_train, X_test, y_train, y_test = train_test_split(X_scaled,
y, test_size=0.25)
```

With our dataset prepared, we are now ready to go ahead and start exploring a few regression models.

Exploring different regression models

There are many types of regression methods that can be used with a given dataset. We can think of regression as falling into four main categories: linear regressions, logistic regressions, ensemble regressions, and finally, boosted regressions. Throughout the next section, we will be exploring examples from each of these categories, starting with linear regression.

Single and multiple linear regression

In many of the datasets you will likely encounter in your career, you oftentimes find that some of the features exhibit some type of correlation vis-à-vis one another. Earlier in this chapter, we discussed the idea of a correlation between two features as a dependence of one feature upon another, which can be calculated using the Pearson correlation metric known as **R2**. Over the last few chapters, we have looked at the idea of correlations in a few different ways, including heats maps and pairplots.

Using the dataset we just prepared, we can take a look at a few correlations using `seaborn`:

```
import seaborn as sns
fig = sns.pairplot(data=df[["Heteroatoms", "MolWt",
"HeavyAtoms"]])
```

This yields the following figure:

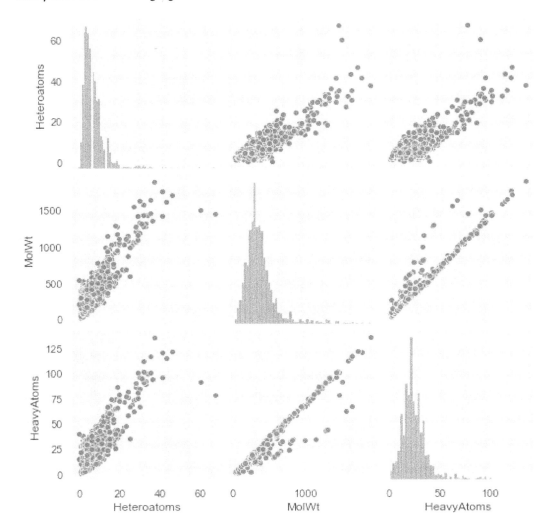

Figure 7.41 – Results of the pairplot function using the toxicity dataset

We can see that there are a few correlations in our dataset already. Using **simple linear regression**, we can take advantage of this correlation in the sense that we can use one variable to predict what the other will most likely be. For example, if X was the independent variable, and Y was the dependent variable, we can define the linear relationship between the two as:

$$y = mX + c$$

In which m is the slope and c is the y intercept. This equation should be familiar to you based on some of the earlier content in this book, as well as your math classes in high school. Using this relationship, our objective will be to optimize this line relative to our data in order to determine the values for m and c using a method known as the least squares method.

Before we can discuss the **least squares method**, let's first discuss the idea of a **loss function**. A loss function within the context of machine learning is a measure of the difference between our calculated and expected values. For example, let's examine the **quadratic loss function**, commonly used to calculate loss within a regression model, which we can define as:

$$L(x) = \sum_{i=1}^{n} (y_i - p_i)^2$$

By now, I would hope you recognize this function from our discussions in the *Measuring success* section. Can you tell me where we last used this?

Now that we have discussed the idea of loss functions, let's take a closer look at the **least squares method**. The main idea behind this mathematical method is to determine a line of best fit for a given set of data, as demonstrated by the correlations we just saw, by minimizing the loss. To fully explain the concepts behind this equation we would need to discuss partial derivatives and what not. For the purposes of simplicity, we will simply define the least squares method as a process for **minimizing** loss in order to determine a line of best fit for a given dataset.

We can divide linear regressions into two main categories: **simple linear regression** and **multiple linear regression**. The main idea here concerns the number of features we will be training the model with. If we are simply training the model based on one feature, we will be developing a simple linear regression. On the other hand, if we train the model using multiple features, we would be training a multiple regression model.

Whether you are training a simple or multiple regression model, the process and desired outcomes are generally the same. Ideally, the output of our models when plotted against the actual values should result in a linear line showing a strong correlation in our data, as depicted in *Figure 7.42*:

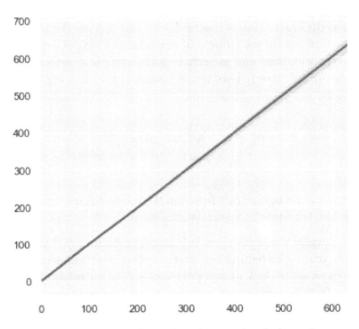

Figure 7.42 – A simple linear line showing the ideal correlation

Let's go ahead and explore the development of a multiple linear regression model. With the data imported in the previous section, we can import the LinearRegression class for sklearn, fit it with our training data, and make a prediction using the test dataset:

```
from sklearn.linear_model import LinearRegression
reg = LinearRegression().fit(X_train, y_train)
y_pred = reg.predict(X_test)
```

Next, we can go ahead and use the actual and predicted values to both calculate the **R2** value and visualize on the graph. In addition, we will also capture the **MSE** metric:

```
p = sns.jointplot(x=y_test, y=y_pred, kind="reg")
p.fig.suptitle(f"Linear Regression, R2 = {round(r2_score(y_
test, y_pred), 3)}, MSE = {round(mean_squared_error(y_test,
y_pred), 2)}")
p.fig.subplots_adjust(top=0.90)
```

The code will then yield the following figure:

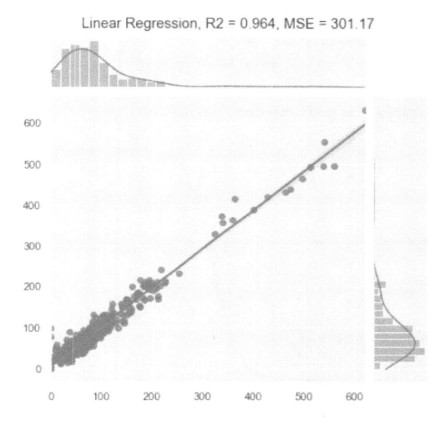

Figure 7.43 – Results of the linear regression model

Immediately, we notice that this simple linear regression model was quite effective in making predictions on our dataset. Notice that this model not only used one feature, but used all of the features to make its prediction. We notice from the graph that the vast majority of our data is localized at the bottom left. This is not ideal for a regression model as we would prefer the values to be equally dispersed across all bounds; however, it is important to remember that in the biotechnology space, you will almost always encounter real-world data in which you will observe items such as these.

If you are following along in **Jupyter Notebooks**, go ahead and reduce the dataset to only one input feature, scale and split the data, train a simple linear regression, and compare the results to the multiple linear regression. Does the correlation of our predictions and actual values increase or decrease?

Logistic regression

Recall that in the linear regression section, we discussed the methodology in which a single linear line can be used to predict a value based on a correlated feature as input. We outlined the linear equation as follows:

$$y_i = \beta x_i + a + \varepsilon_i$$

In some instances, the relationship between the data and the desired output may not be best represented by a linear model, but rather a non-linear one:

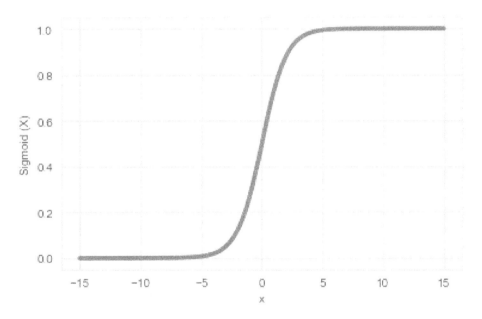

Figure 7.44 – A simple sigmoid curve

Although known as **logistic regression**, this regression is mostly used as a **binary classification** algorithm. However, the main focus here is that the word *logistic* is referring to the **logistic function**, also known as the **Sigmoid** function, represented as:

$$y = \frac{1}{1 + e^{-x}}$$

With this in mind, we will want to use this function to make predictions in our dataset. If we wanted to determine whether or not a compound was toxic given a specific input value, we could calculate the weighted sum of inputs such that:

$$x = \Theta * input + b$$

This would allow us to calculate the probability of toxicity such that:

$$probability\ of\ toxicity(x) = \frac{1}{1 + e^{-x}}$$

Using this probability, a final prediction can be made, and an output value assigned. Go ahead and implement this model using the protein classification dataset in the previous section and compare the results that you find to those of the other classifiers.

Decision trees and random forest regression

Similar to its classification counterpart, **Decision Tree Regressions (DTRs)** are commonly used machine learning models implementing nearly the same internal mechanisms that decision tree classifiers use. The only difference between the models is the fact that regressors output continuous numerical values, whereas classifiers output discrete classes.

Similarly, another model known as **Random Forest Regressors (RFRs)** also exists and operates similarly to its classification counterparts. This model is also an ensemble method in which each tree is trained as a separate model and subsequently averaged.

Let's go ahead and implement an RFR using this dataset. Just as we have previously done, we will first create an instance of the model, fit it to our data, make a prediction, and visualize the results:

```
from sklearn.ensemble import RandomForestRegressor
reg = RandomForestRegressor().fit(X_train, y_train)
y_pred = reg.predict(X_test)
p = sns.jointplot(x=y_test, y=y_pred, kind="reg")
p.fig.suptitle(f"RandomForestRegressor Regression, R2 =
{round(r2_score(y_test, y_pred), 3)}, MSE = {round(mean_
squared_error(y_test, y_pred), 2)}")
# p.ax_joint.collections[0].set_alpha(0)
# p.fig.tight_layout()
p.fig.subplots_adjust(top=0.90)
```

With the model developed, we can visualize the results using the following plot:

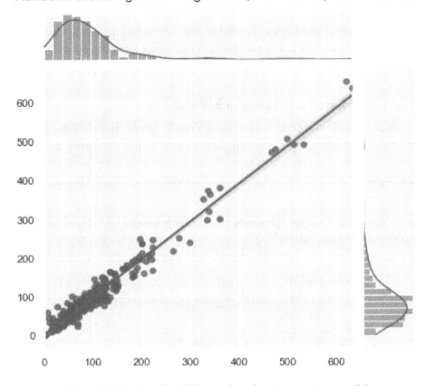

RandomForestRegressor Regression, R2 = 0.963, MSE = 280.88

Figure 7.45 – Results of the random forest regression model

Notice that while the **R2** remained relatively the same, the **MSE** did decrease substantially relative to the first linear model. We can improve these scores by tuning the model's parameters. We can demonstrate this using the max_depth parameter:

```
for i in range(1,10):
    reg = RandomForestRegressor(max_depth=i)
                     .fit(X_train, y_train)
    y_pred = reg.predict(X_test)
    print("depth =", i,
          "score=", r2_score(y_test, y_pred),
          "mse = ", mean_squared_error(y_test, y_pred))
```

The output for this code is shown below, indicating that a `max_depth` of 8 would likely be optimal given that it results in an **R2** of 0.967 and an **MSE** of 248.133:

```
depth = 1 score= 0.6829573377527721 mse =  2408.705147781684
depth = 2 score= 0.8551988639595177 mse =  1100.1145376244872
depth = 3 score= 0.9251880659069978 mse =  568.377421158544
depth = 4 score= 0.9435405882402688 mse =  428.94566549010085
depth = 5 score= 0.9564044343774294 mse =  331.21366882055423
depth = 6 score= 0.9614323533531397 mse =  293.0144743223112
depth = 7 score= 0.9638838635702021 mse =  274.3893301924249
depth = 8 score= 0.9673397474729089 mse =  248.13354086873196
depth = 9 score= 0.9664514650632033 mse =  254.88219228925627
```

Figure 7.46 – Results of the random forest regression model with differing max_depth

Similar to classification, decision trees for regression tend to be excellent methods for allowing you to develop models while trying to avoid overfitting your data. Another great benefit of **DTR** models, when using the **sklearn** API, is gaining insights into feature importance directly from the model. Let's go ahead and demonstrate this:

```
features = X.columns
importances = reg.feature_importances_
indices = np.argsort(importances)[-9:]
plt.title('Feature Importances')
plt.barh(range(len(indices)), importances[indices],
                    color='royalblue',
                    align='center')
plt.yticks(range(len(indices)), [features[i] for i in indices])
plt.xlabel('Relative Importance')
plt.show()
```

With that complete, this yields the following diagram:

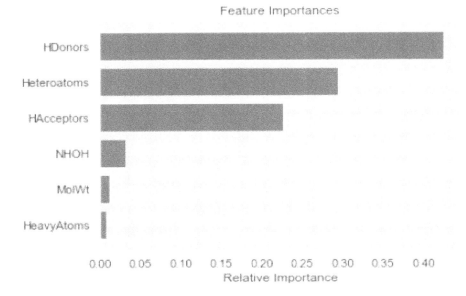

Figure 7.47 – Feature importance of the random forest regression model

Looking at this figure, we can see that the top three features that had the biggest impact on the development of this model were HDonors, Heteroatoms, and HAcceptors. Although this example of feature importance was developed using the RFR model, we can theoretically use this with many other models. One library in particular that has gained a great deal of importance in the field concerning the idea of feature importance is the SHAP library. It is highly recommended that you take a glance at this library and the many wonderful features (no pun intended) that it has to offer.

XGBoost regression

Similar to the **XGBoost** classification model we investigated in the previous section, we also have regression implementation, which allows us to predict a value rather than a category. We can go ahead and implement this easily, just as we did in the previous section:

```
import xgboost as xg
reg = xg.XGBRegressor(objective ='reg:linear',n_estimators =
1000).fit(X_train, y_train)
y_pred = reg.predict(X_test)


p = sns.jointplot(x=y_test, y=y_pred, kind="reg")
```

```
p.fig.suptitle(f"xgboost Regression, R2 = {round(r2_score(y_
test, y_pred), 3)}, MSE = {round(mean_squared_error(y_test,
y_pred), 2)}")
p.fig.subplots_adjust(top=0.90)
```

With the code completed, this yields the following figure:

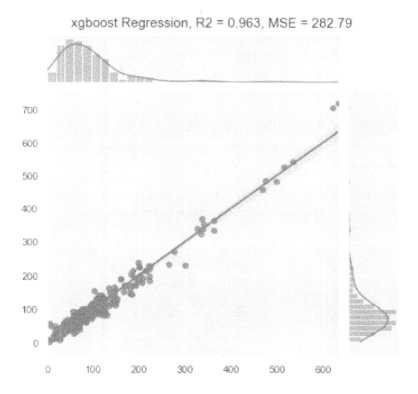

Figure 7.48 – Results of the XGBoost regression model

You will notice that this implementation of the model gave us a very respectable **R2** value in the context of our actual and predicted values and managed to yield an **MSE** of 282.79, which is slightly less than some of the other models we've attempted in this chapter. With the model completed, let's now move on to see how we could use some of the automated machine learning functionality provided in AWS in the following tutorial.

Tutorial: Regression for property prediction

Over the course of this chapter, we have reviewed some of the most common (and popular) regression models as they relate to the prediction of the TPSA feature using the toxicity dataset. In the previous section pertaining to classification, we created a GCP instance and used the auto-sklearn library to automatically identify one of the top machine learning models for a given dataset. In this tutorial, we will create an **AWS SageMaker** notebook, query our dataset from RDS, and run the auto-sklearn library in a similar manner. In addition, we will also explore an even more powerful automated machine learning method using **AWS Autopilot**. In one of the earlier chapters, we used **AWS RDS** to launch a relation database to host our toxicity dataset. Using that same **AWS** account, we will now go ahead and get started.

Creating a SageMaker notebook in AWS

Similar to the creation of a notebook in **GCP**, we can create a **SageMaker** notebook in **AWS** in a few simple steps:

1.   Navigate to the AWS Management Console on the front page. Click on the **Services** drop-down menu and select **Amazon SageMaker** under the **Machine Learning** section:

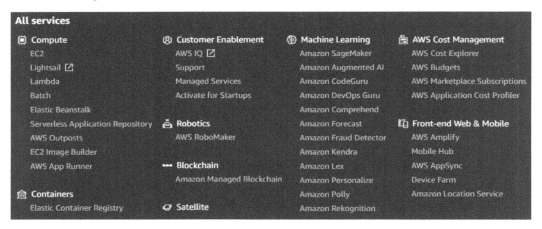

Figure 7.49 – The list of services provided by AWS

2.   On the left-hand side of the page, click the **Notebook** drop-down menu and then select the **Notebook instances** button:

▼ **Notebook**

Notebook instances

Lifecycle configurations

Git repositories

Figure 7.50 – A screenshot of the notebook menu

3. Within the notebook instances menu, click on the orange button called **Create notebook instance**:

Figure 7.51 – A screenshot of the Create notebook instance button

4. Let's now go ahead and give the notebook instance a name, such as `biotech-machine-learning`. We can leave the instance type as the default selection of `ml.t2.medium`. This is a medium-tier instance and is more than enough for the purposes of our demo today:

Notebook instance settings

Notebook instance name

biotech-machine-learning

Maximum of 63 alphanumeric characters. Can include hyphens (-), but not spaces. Must be unique within your account in an AWS Region.

Notebook instance type

ml.t2.medium ▼

Elastic Inference Learn more ☑

none ▼

▶ Additional configuration

Figure 7.52 – A screenshot of the notebook instance settings

5.   Under the **Permissions and encryption** section, select the **Create a new role** option for the IAM role section. The main idea behind IAM roles is the concept of granting certain users or roles the ability to interact with specific AWS resources. For example, we could allow this role to also have access to some but not all S3 buckets. For the purposes of this tutorial, let's go ahead and grant this role access to any S3 bucket. Go ahead and click on **Create role**:

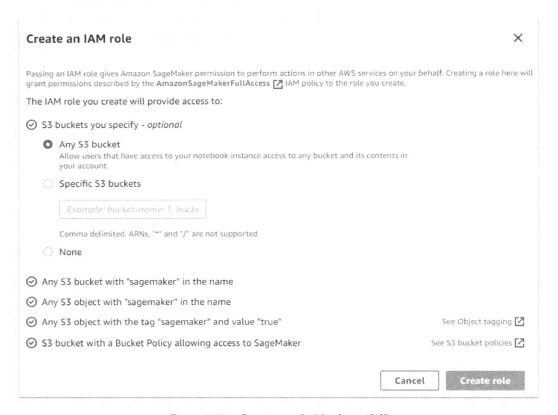

Figure 7.53 – Creating an IAM role in AWS

6.   Leaving all the other options as they are, go ahead and click on **Create notebook instance**. You will be redirected back to the **Notebook instance** menu where you will see your newly created instance in a **Pending** state. Within a few moments, you will notice that status change to **InService**. Click on the **Open JupyterLab** button to the right of the status:

Figure 7.54 – The options to open a Jupyter notebook or Jupyer lab in AWS

7. Once again, you will find yourself in the familiar Jupyter environment you have been working in.

AWS SageMaker is a great resource for you to use at a very low cost. Within this space, you will be able to run all of the Python commands and libraries you have learned throughout this book. You can create directories to organize your files and scripts and access them anywhere in the world without having to have your laptop with you. In addition, you will also have access to almost 100 SageMaker sample notebooks and starter code for you to use.

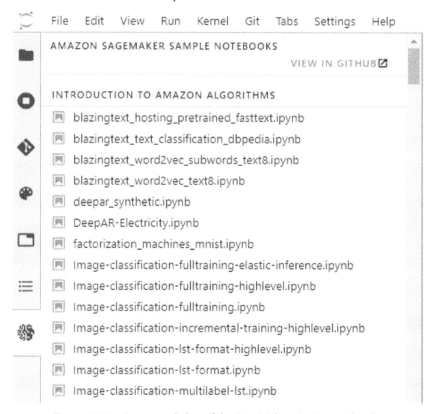

Figure 7.55 – An example list of the SageMaker starter notebooks

With this final step complete, we now have a fully working notebook instance. In the following section, we will use SageMaker to import data and start running our models.

Creating a notebook and importing data in AWS

Given that we are once again working in our familiar Jupyter space, we can easily create a notebook by selecting one of the many options on the right-hand side and start running our code. Let's go ahead and get started:

1. We can begin by selecting the **conda_python3** option on the right-hand side, creating a new notebook for us in our current directory:

Figure 7.56 – A screenshot of conda_python3

2. With the notebook prepared, we will need to install a few libraries to get started. Go ahead and install mysql-connector and pymysql using pip:

```
pip install mysql-connector pymysql
```

3. Next, we will need to import a few things:

```
import pandas as pd
import mysql.connector
from sqlalchemy import create_engine
import sys
import seaborn as sns
```

4. Now, we can define some of the items we will need to query our data, as we did previously in *Chapter 3*, *Getting Started with SQL and Relational Databases*:

```
ENDPOINT="toxicitydataset.xxxxxx.us-east-2.rds.amazonaws.
com"
PORT="3306"
USR="admin"
```

```
DBNAME="toxicity_db_tutorial"
PASSWORD = "xxxxxxxxxxxxxxxxxx"
```

5. Next, we can go ahead and create a connection to our **RDS** instance:

```
db_connection_str = 'mysql+pymysql://{USR}:{PASSWORD}@
{ENDPOINT}:{PORT}/{DBNAME}'.format(USR=USR,
PASSWORD=PASSWORD, ENDPOINT=ENDPOINT, PORT=PORT,
DBNAME=DBNAME)
db_connection = create_engine(db_connection_str)
```

6. Finally, we can go ahead and query our data using a basic **SQL** SELECT statement:

```
df = pd.read_sql('SELECT * FROM dataset_toxicity_sd',
con=db_connection)
```

With that complete, we are now able to query our data directly from **AWS RDS**. As you begin to explore new models in the realm of data science, you will need a place to store and organize your data. Selecting a platform such as **AWS RDS**, **AWS S3**, or even **GCP BigQuery** will help you organize your data and studies.

Running auto-sklearn using the toxicity dataset

Now that we have our data in a working notebook, let's go ahead and use the auto-sklean library to identify a model best suited for our given dataset:

1. We can begin by installing the auto-sklearn library in our **SageMaker** instance:

```
pip install auto-sklearn
```

2. Next, we can isolate our input features and output values and scale them accordingly:

```
X = df[["Heteroatoms", "MolWt", "HeavyAtoms", "NHOH",
"HAcceptors", "HDonors"]]
y = df.TPSA.values.ravel()
from sklearn.preprocessing import MinMaxScaler
scaler = MinMaxScaler()
X_scaled = scaler.fit_transform(X)
```

3. With the data scaled, we can now go ahead and separate our training and test datasets:

```
from sklearn.model_selection import train_test_split
X_train, X_test, y_train, y_test = train_test_split(X_
scaled, y, test_size=0.25)
```

4. Finally, we can import the regression implementation of `sklearn`, adjust the parameters, and fit the model to our dataset:

```
import autosklearn.regression
automl = autosklearn.regression.AutoSklearnRegressor(
    time_left_for_this_task=120,
    per_run_time_limit=30,
    tmp_folder='/tmp/autosklearn_regression_example_
tmp2')
automl.fit(X_train, y_train, dataset_name='dataset_
toxicity')
```

5. Once the model is done, we can take a look at the top-performing candidate model using the `get_models_with_weights()` function:

```
automl.get_models_with_weights()[0]
```

6. Lastly, we can go ahead and get a sense of the model's performance using the **R2** and **MSE** metrics, as we have done previously:

```
from sklearn.metrics import r2_score, mean_squared_error
predictions = automl.predict(X_test)
p = sns.jointplot(x=y_test, y=predictions, kind="reg")
p.fig.suptitle(f"automl, R2 = {round(r2_score(y_test,
predictions), 3)}, MSE = {round(mean_squared_error(y_
test, predictions), 2)}")
# p.ax_joint.collections[0].set_alpha(0)
# p.fig.tight_layout()
p.fig.subplots_adjust(top=0.90)
```

Upon plotting the output, this yields the following results:

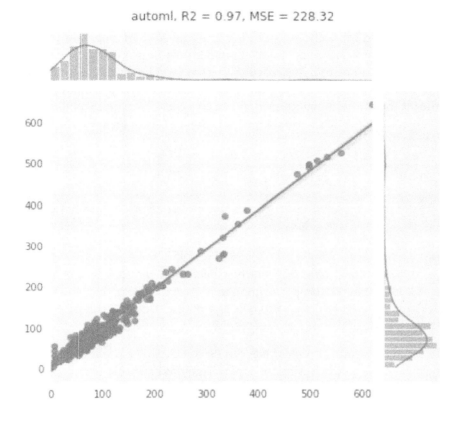

automl, R2 = 0.97, MSE = 228.32

Figure 7.57 – Results of the AutoML model showing the correlation of 0.97

We can see from the figure above that the actual results and predicted results line up quite nicely, giving us an R2 value of approximately 0.97, showing a strong correlation. In the following section, we will explore the process of automating parts of the model development process using AWS Autopilot.

Automated regression using AWS Autopilot

Many different tools and applications can be found in the **AWS Management Console**, offering solutions to many data science and computer science problems any developer will likely encounter. There is one tool in particular that has stood out and begun to grow in popularity among the data science community known as **AWS Autopilot**. The purpose of **AWS Autopilot** is to help automate some of the tasks generally undertaken in any given data science project. We can see a visual representation of this in *Figure 7.58*:

Figure 7.58 – The Autopilot pipeline

Users are able to load in their datasets, identify a few parameters, and let the model take it from there as it identifies the top-performing models, optimizes a few parameters, and even generates sample code for the user to take and optimize even further. Let's go ahead and demonstrate the use of this model using the same dataset:

1. We can begin by creating a SageMaker Studio instance by navigating to the SageMaker console and selecting the **Open SageMaker Studio** button on the right. Using the quick start option, the default settings, and a new **IAM role**, click the **Submit** button. After a few moments, the instance will provision. Click on the **Open Studio** button:

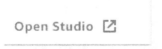

Figure 7.59 – The Open Studio option in AWS

2. While the instance is provisioning, let's upload our dataset to **S3**. Using the **AWS Management Console**, select the **S3** option under the storage section. Here, you can create buckets, which act as online storage spaces. Create a new bucket called `biotech-machine-learning` while keeping all the other options at their default values.

Figure 7.60 – Creating a bucket in AWS

3. Once created, open the bucket and click on the **Upload** button. Then, upload the CSV file of the reduced and cleaned proteins dataset.

Figure 7.61 – Uploading files in AWS

4. With the dataset uploaded, let's now head back to SageMaker. Using the navigation pane on the left, select the **SageMaker Components and Registries** tab. Using the drop-down menu, select **Experiments and trials,** and then click the **Create Autopilot Experiment** button:

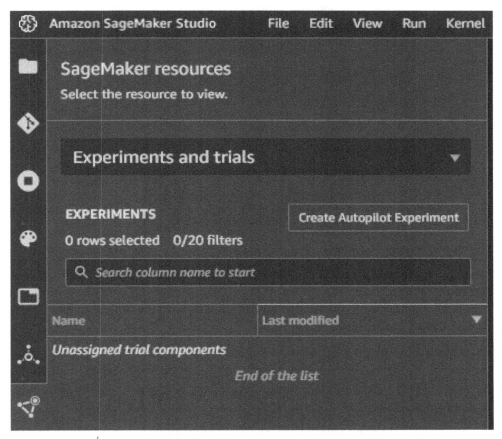

Figure 7.62 – Creating SageMaker resources in AWS SageMaker Studio

5. Let's move on and give the experiment a name, such as `dataset-pdb-nodups-cleaned`.

6. In the **CONNECT YOUR DATA** section, select the S3 bucket name you created earlier, as well as the dataset filename:

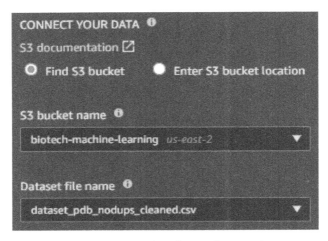

Figure 7.63 – Connecting data to the experiment

7. Next, select the target column, which in our case, is the classification column:

Figure 7.64 – Selecting a target column for the model training process

8. Finally, you can now go ahead and disable the **Auto deploy** option and click **Create Experiment**. Similar to GCP's **AutoML**, the application will identify a set of models deemed to be the best fit for your given dataset. You have the option to select between **Pilot** or **Complete**.

   A complete experiment will train and tune the model while allowing users to view the details and statistics in real time. It will go through various phases, such as preprocessing, candidate definition generation, feature engineering, model tuning, and report generation.

9. Upon completing the process, a dashboard with all the trained models and their associated metrics will be presented, as depicted in *Figure 7.65*. Users can explore the models and deploy them in a few simple clicks.

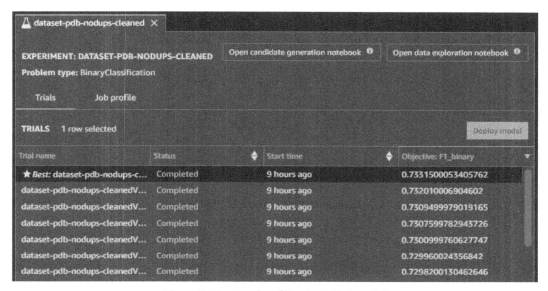

Figure 7.65 – Results of the Autopilot model

AWS Autopilot is a robust and useful tool that every data scientist can utilize when facing a difficult dataset. It not only assists in identifying the best model for a given dataset, but can also help preprocess the data, tune the model, and provide sample code for users to use.

Summary

Congratulations! We finally made it to the end of a very dense, yet very informative chapter. In this chapter, we learned quite a few different things. In the first half of this chapter, we explored the realm of classification and demonstrated the application of a number of models using the single-cell RNA dataset – a classical application in the field of biotechnology and life sciences. We learned about a number of different models, including KNNs, SVMs, decision trees, random forests, and XGBoost. We then moved our data and code to GCP, where we stored our data in BigQuery, and provisioned a notebook instance to run our code in. In addition, we learned how to automate some of the manual and labor-intensive parts of the model development process as it pertains to the protein classification dataset using auto-sklearn. Finally, we took advantage of GCP's AutoML application to develop a classification model for our dataset.

In the second half of this chapter, we explored the realm of regression as it pertains to the toxicity dataset. We explored the idea of correlation within data and learned a few important regression models too. Some of the models we looked at included simple and multiple linear regression, logistic regression, decision tree regressors, and an XGBoost regressor as well. We then moved our code to AWS's SageMaker platform and used the previously provisioned RDS to query our data and run auto-sklearn for regression as well. Finally, we implemented AWS Autopilot's automated machine learning model for the toxicity dataset.

So far, we have spent much of our time developing machine learning models using the `sklearn` library. However, not every dataset can be classified or regressed using machine learning – sometimes, a more powerful set of models will be needed. For datasets such as those, we can turn to the field of deep learning, which will be our focus for the next chapter.

8
Understanding Deep Learning

Throughout this book, we have examined the many tools and methods within the fields of supervised and unsupervised machine learning. Within the field of unsupervised learning, we explored **clustering** and **dimensionality reduction**, while within the field of supervised learning, we explored **classification** and **regression**. Within all of these fields, we explored many of the most popular algorithms for developing powerful predictive models for our datasets. However, as we have seen with some of the data we have worked with, there are numerous limitations when it comes to these models' performance that cannot be overcome by additional tuning and hyperparameter optimization. In cases such as these, data scientists often turn to the field of **deep learning**.

If you recall our overarching diagram of the artificial intelligence space that we saw in *Chapter 5, Introduction to Machine Learning*, we noted that the overall space is known as **Artificial Intelligence** (**AI**). Within the AI space, we defined machine learning as the ability to develop models to learn or generalize from data and make predictions. We will now explore a subset of machine learning known as **deep learning**, which focuses on developing models and extracting patterns within data using deep neural networks.

Throughout this chapter, we will explore the ideas of neural networks and deep learning as they relate to the field of biotechnology. In particular, we will be covering the following topics:

- Understanding the field of deep learning
- Exploring the types of deep learning models
- Selecting an activation function
- Measuring progress with loss
- Developing models with the Keras library
- Tutorial – protein sequence classification via LSTMs using Keras and MLflow
- Tutorial – anomaly detection using AWS Lookout for Vision

With these sections in mind, let's get started!

Understanding the field of deep learning

As we mentioned in the introduction, deep learning is a subset or branch of the machine learning space that focuses on developing models using neural networks. The idea behind using neural networks for deep learning derives from neural networks found in the human brain. Let's learn more about this.

Neural networks

Similar to machine learning, the idea behind developing deep learning models is not to explicitly define the steps in which a decision or prediction is made. The main idea here is to generalize from the data. Deep learning makes this possible by drawing a parallel between the dendrites, cell body, and synapses of the human brain, which, within the context of deep learning, act as inputs, nodes, and outputs for a given model, as shown in the following diagram:

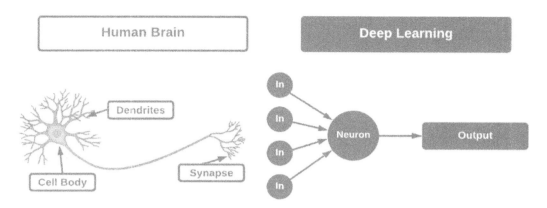

Figure 8.1 – Comparison between the human brain and a neural network

Some of the biggest benefits behind such an implementation revolve around the idea of feature engineering. Earlier in this book, we saw how features can be created or summarized using various methods such as basic mathematical operations (x2) or through complex algorithms such as **Principal Component Analysis** (**PCA**). Manually engineered features can be very time-consuming and not feasible in practice, which is where the field of deep learning can come in, with the ability to learn the many underlying features in a given dataset directly from the data.

Within the field of **biotechnology**, most applications, ranging from the early stages of therapeutic discovery all the way downstream to manufacturing, are generally data-rich processes. However, much of the data that's been collected will have little to no use on its own, or perhaps the data that's been collected is for different batches of a particular molecule. Perhaps the data is extensive for some molecules and less extensive for others. In many of these cases, using deep learning models can come to your aid when it comes to features that are relative to the traditional machine learning models we have discussed so far.

We can think of features at three different levels:

- **Low-level features**, such as individual amino acids, a protein, or the elements of a small molecule.

- **Mid-level features**, such as the amino acid sequences of a protein and the functional groups of a small molecule.

- **High-level features**, such as the overall structure or the classification of a protein or the geometric shape of a small molecule.

The following diagram shows a graphical representation of these features:

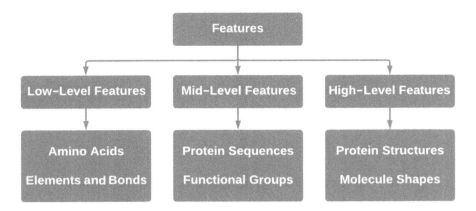

Figure 8.2 – The three types of features and some associated examples

In many instances, architecting a robust deep learning model can unlock a more powerful predictive model relative to its machine learning counterpart. In many of the machine learning models we have explored, we attempted to improve the model's performance not only by tuning and adjusting the hyperparameters but also by making a conscious decision to use datasets with a sufficient amount of data. Increasing the size of the datasets will likely not lead to any significant improvement in our machine learning models. However, this is not always the case with deep learning models, which tend to improve in performance when more data is made available. We can see a visual depiction of this in the following diagram:

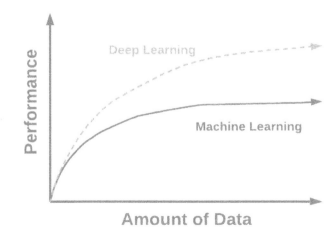

Figure 8.3 – A graphical representation of machine learning versus deep learning

Using neural networks within the context of machine learning has seen a major surge in recent years, which can be attributed to the increased use of big data within most industries, the decreased expense of computational hardware such as CPUs and GPUs, and the growing community that supports much of the open source software and packages that are available today. Two of the most common packages out there for developing deep learning models are TensorFlow and Keras – we will explore these two later in this chapter. Before we do, let's go ahead and talk about the architecture behind a deep learning model.

The perceptron

One of the most important building blocks of any deep learning model is the perceptron. A perceptron is an algorithm that's used for developing supervised binary classifiers, first invented in 1958 by Frank Rosenblatt, who is sometimes called the father of deep learning. A perceptron generally consists of four major parts:

- The **input** values, which are generally taken from a given dataset.
- The **weights**, which are values by which the input values are multiplied.
- The **net sum**, which is the sum of all the values from each of the inputs.
- The **activation function**, which maps a resulting value to an output.

The following diagram shows a graphical representation of these four parts of a perceptron:

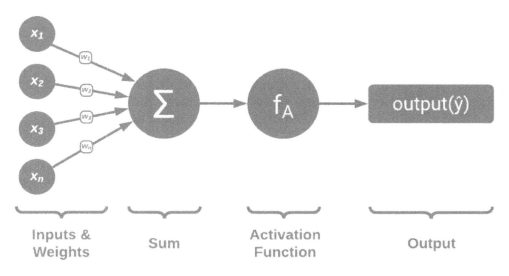

Figure 8.4 – A graphical representation of a perceptron

There are three main steps that a perceptron takes to arrive at a predicted output from a given set of input values:

1. The **input values** (x_1, x_2, and so on) are multiplied by their respective weights (w_1, w_2, and so on). These weights are determined in the training process for this model so that a different weight is assigned to each of the input values.

2. All the values from each of the calculations are summed together in a value known as the **weighted sum**.

3. The weighted sum is then applied to the **activation function** to map the value to a given output. The specific activation function that's used is dependent on the given situation. For example, within the context of a unit step activation function, values would either be mapped to 0 or 1.

When viewed from a mathematical perspective, we can define the output value, $\hat{y}$, as follows:

$$\hat{y} = g(w_0 + \sum_{i=1}^{m} x_i w_i)$$

In this equation, g is the activation function, w_0 is the bias, and the final components are the sum of the **linear combination** of input values:

$$\hat{y} = g(w_0 + X^T W)$$

So, in this equation, $X = \begin{bmatrix} x_1 \\ \cdots \\ x_m \end{bmatrix}$ and $W = \begin{bmatrix} w_1 \\ \cdots \\ w_m \end{bmatrix}$ account for the final output value.

A perceptron is one of the simplest deep learning building blocks out there and can be expanded quite drastically by increasing the number of **hidden layers**. Hidden layers are the layers that lay in-between the input and output layers. Models with very few hidden layers are generally referred to as **neural networks** or multilayer perceptrons, whereas models with many hidden layers are referred to as **deep neural networks**.

Each of these layers consists of several **nodes**, and the flow of data is similar to that of the perceptron we saw previously. The number of input nodes (x_1, x_2, x_3) generally corresponds to the number of **features** in a given dataset, whereas the number of output nodes generally corresponds to the number of outputs.

The following diagram is a graphical representation of the difference between neural networks and deep learning:

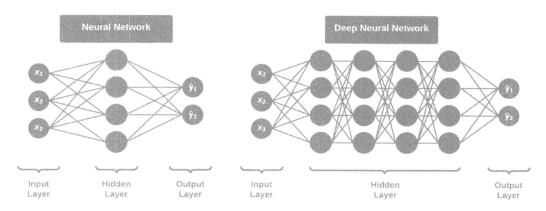

Figure 8.5 – Difference between neural networks and deep learning

In the previous diagram, we can see the neural network or multilayer perceptron (on the left) consisting of an input layer, a single hidden layer with four nodes, and an output layer with four nodes. Similar to the single perceptron we saw earlier, the idea here is that each of the nodes within the hidden layer will intake the input nodes, multiplied by some value, and then pass them through an **activation function** to yield output. On the right, we can see a similar model, but the values are passed through several hidden layers before determining a final output value.

Exploring the different types of deep learning models

There are many different types of neural networks and deep learning architectures out there that differ in function, shape, data flow, and much more. There are three types of neural networks that have gained a great deal of popularity in recent years, given their promise and robustness with various types of data. First, we will explore the simplest of these architectures, known as a multilayer perceptron.

Multilayer perceptron

A **Multilayer Perceptron** (**MLP**) is one of the most basic types of **Artificial Neural Networks** (**ANNs**). This type of network is simply composed of layers in which data flows in a forward manner, as shown in the previous diagram. Data flows from the input layer to one or more hidden layers, and then finally to an output layer in which a prediction is produced. In essence, each layer attempts to learn and calculate certain weights. ANNs and MLPs come in many different shapes and sizes: they can have a different number of nodes in each layer, a different number of inputs, or even a different number of outputs. We can see a visual depiction of this in the following diagram:

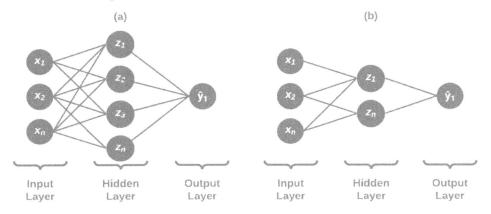

Figure 8.6 – Two examples of MLPs

MLP models are generally very versatile but are most commonly used for structured tabular data, such as the structured protein classification dataset we have been working with. In addition, they can be used for image data or even text data. MLPs, however, generally tend to suffer when it comes to sequential data such as protein sequences and time-series datasets.

Convolutional neural networks

Convolutional Neural Networks (**CNNs**) are deep learning algorithms that are commonly used for processing and analyzing image data. CNNs can take in images as input data and restructure them to determine the importance through weights and biases, allowing it to distinguish between the features of one image relative to another. Similar to our earlier discussion of how deep learning is similar to neurons in the brain, CNNs are also analogous to the connectivity of neurons in the human brain and the visual cortex when it comes to the sensitivity of regions, similar to the concept of receptive fields. One of the biggest areas of success for CNN models is their ability to capture spatial dependencies, as well as temporal dependencies, in images through the use of filters. We can see a visual representation of this in the following diagram:

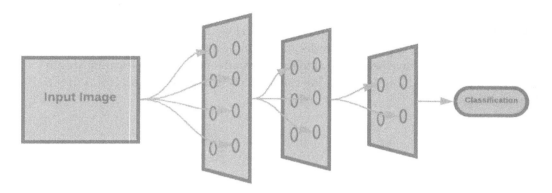

Figure 8.7 – A representation of a CNN

Let's take, for example, the idea of creating an image classification model. We could use an ANN and convert a 2D image of pixels by flattening it. An image with a 4x4 matrix of pixels would now become a 1x16 vector instead. This change would cause two main drawbacks:

- The spatial features of the image would be lost, and thereby reduce the robustness of any trained model.

- The number of input features would increase quite drastically as the image size grows.

CNNs can overcome this by extracting high-level features from the images, allowing them to be quite effective with image-based datasets.

Recurrent neural networks

Recurrent Neural Networks (**RNNs**) are commonly used algorithms that are generally applied to sequence-based datasets. They are quite similar in architecture to the ANNs we discussed earlier, but RNNs can remember their input using internal memory, making them quite effective with sequential datasets in which previous data is of great importance.

Take, for example, a protein sequence consisting of various amino acids. To predict the class of the protein, or its general structure, the model would not only need to know which amino acids were used but the order in which they were used as well. RNNs and their many derivatives have been central to the many advances in deep learning within the field of biology and biotechnology. We can see a visual representation of this in the following diagram:

Recurrent Neural   Artificial Neural
Network   Network

Figure 8.8 – A representation of an ANN node versus an RNN node

There are several advantages when it comes to using RNNs as predictor models, with the main benefits being as follows:

- Their ability to capture the dependency between data points such as words in a sentence
- Their ability to share parameters across time steps, thus decreasing the overall computational cost

Because of this, RNNs have become increasingly popular architectures for developing models that solve problems related to scientific sequence data such as proteins and DNA, as well as text and time-series data.

Long short-term memory

Long Short-Term Memory (**LSTM**) models are a type of RNN designed with the capability of learning long-term dependencies when handling sequence-based problems. Commonly used with text-based data for classification, translation, and recognition, LSTMs have gained an unprecedented surge in popularity over the years. We can depict the structure of a standard RNN as we did previously, but structured slightly differently:

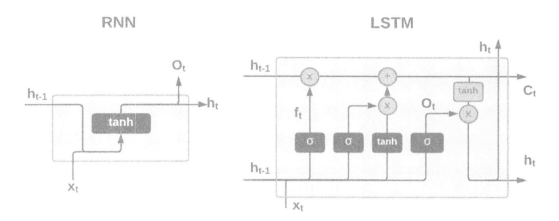

Figure 8.9 – The inner workings of an RNN versus an LSTM

In the preceding diagram, X_t is an input vector, h_t is a hidden layer vector, and o_t is an output vector. On the other hand, and using some of the same elements, an LSTM can be structured quite similarly. Without diving into too much detail, the core idea behind an LSTM is the cell state (the top horizontal line). This state operates similarly to a conveyor belt in which data flows linearly through it. Gates within the cell are methods that optionally allow information to be added to the state. An LSTM has three gates, all leading to the cell state.

Although LSTM models and their associated diagrams can be quite intimidating at first, they have proven their worth time and time again in various areas. Most recently, LSTM models have been used as generative models for antibody design, as well as classification models for protein sequence-structure classification. Now that we have explored several common deep learning architectures, let's go ahead and explore their main components: activation functions.

Selecting an activation function

Recall that, in the previous section, we used an activation function to map a value to a particular output, depending on the value. We will define an activation function as a mathematical function that defines the output of an individual node using an input value. Using the analogy of the human brain, these functions simply act as gatekeepers, deciding what will be *fired off* to the next neuron. There are several features that an activation function should have to allow the model to learn most effectively from it:

- The avoidance of a vanishing gradient
- A low computational expense

Artificial neural networks are trained using a process known as gradient descent. For this example, let's assume that there is a two-layer neural network:

$$First\ Layer = f_1(x)$$

$$Second\ Layer = f_2(x)$$

The overall network can be represented as follows:

$$o(x) = f_2(f_1(x))$$

When the weights are calculated in a step known as a backward pass, the result becomes as follows:

$$o^{'}(x) = f_2(x) * f_1{'}(x)$$

$$f_1(x) = Activation(w_1 * x_1 + b_1)$$

Upon determining the derivative, the function becomes as follows:

$$f_1^{'(x)} = Activation(w_1 * x_1 + b_1) * x_1$$

If this process were to continue through many layers during the backpropagation step, there would be a considerable reduction in the value of the gradient for the initial layers, thus halting the model's ability to learn. This is the concept of **vanishing gradients**.

On the other hand, **computational expense** is also a feature that must be considered before designing and deploying any given model. The activation functions that are applied from one layer to another must be calculated many times, so the expense of the calculation should be kept to a minimum to avoid longer training periods. The flow of information from an input layer to an output layer is called **forward propagation**.

There are many different types of activation functions out there that are commonly used for various purposes. Although this is not a hard rule, some activation functions are generally used with specific deep learning layers. For example, **ReLU** activation functions are commonly used with **CNNs** and **MLPs**. On the other hand, the `sigmoid` and **Tanh** activation functions are commonly used with **RNNs**. Let's take a moment and look at the three most common activation functions you will likely encounter in your journey:

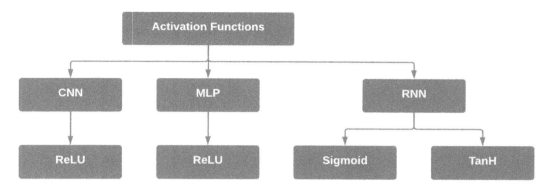

Figure 8.10 – Various types of activation functions by model type

With some of these types now in mind, let's go ahead and explore them in a little more detail. sigmoid functions are probably some of the most commonly used functions within the field of deep learning. It is a non-linear activation function that is also sometimes referred to as a logistic function (remember logistic regression? Hint hint). sigmoid functions are unique in the sense that they can map values to either a 0 or a 1. Using the numpy library, we can easily put together a Python function to calculate it:

```
import numpy as np
def sigmoid_function(x):
    return 1 / (1 + np.exp(-x))
```

Using this function alongside the numpy library, we can generate some data and plot our sigmoid function accordingly:

```
x1 = np.linspace(-10, 10, 100)
y1 = [sigmoid_function(i) for i in x1]
plt.plot(x1,y1)
```

In return, we yield the following diagram, showing the curved nature of a `sigmoid` function. Notice how the upper and lower ranges are 1 and 0, respectively:

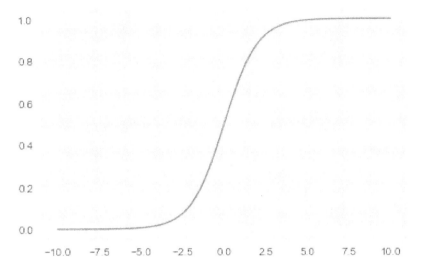

Figure 8.11 – A simple sigmoid function

One of the biggest issues with a `sigmoid` activation function is that the outputs can **saturate** in the sense that values greater than 1.0 are mapped to one, and values that are smaller than 0 are mapped to 0. This can cause some models to fail to generalize or learn from the data and is related to the vanishing gradients issue we discussed earlier in this chapter.

On the other hand, another common activation function is **Tanh**, which is very similar to the `sigmoid` function. The Tanh function is symmetric in the sense that it passes through the point (0, 0) and it ranges to the values of 1 and -1, unlike its `sigmoid` counterpart, making it a slightly better function. Instead of defining our functions in Python, as we did previously, we can take advantage of the optimized functions in the numpy library:

```
x2 = np.linspace(-5, 5, 100)
y2 = np.tanh(x2)
plt.plot(x2, y2)
```

Upon executing this code, we retrieve the following diagram. Notice how the center of the diagram is the point (0, 0), while the upper and lower values are 1.00 and -1.00, respectively:

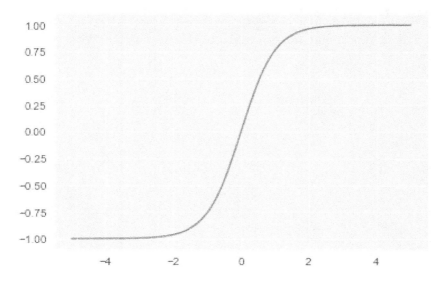

Figure 8.12 – A simple Tanh function

Similar to its `sigmoid` counterpart, the **Tanh** activation function also suffers from the issue of **vanishing gradients** and saturation. However, given the fact that the range is -1 to 1, the gradient is somewhat stronger than `sigmoid`, making it a slightly better function to use.

Finally, yet another commonly used activation function is the **Rectified Linear Unit (ReLU)**. **ReLU** activation functions were specifically developed to avoid saturation when handling larger numbers. The non-linear nature of this function allows it to learn the patterns within the data, whereas the linear nature of the function allows it to be easily interpretable relative to the other functions we have seen so far. Let's go ahead and explore this in Python:

```python
def relu_function(x):
    return np.array([0, x]).max()
x3 = np.linspace(-5, 5, 100)
y3 = [relu_function(i) for i in x3]
plt.plot(x3, y3)
```

Executing this code yields the following diagram. Notice how the **ReLU** function takes advantage of both the linear and non-linear nature of activation functions, giving it the best of both worlds:

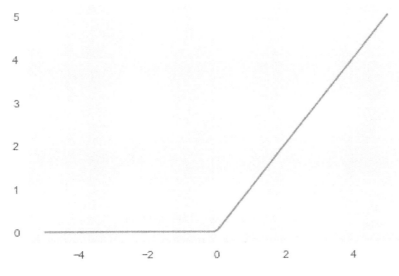

Figure 8.13 – A simple ReLU function

The **ReLU** activation function has become one of the most popular, if not **the** most popular, activation function among data scientists because of its ease of implementation, and its robust speed within the model development and training process. **ReLU** activation functions do, however, have their downsides. For example, the function cannot be differentiable when x = 0 (at point 0, 0), so **gradient descent** cannot be computed for that value.

Yet another activation function worth mentioning is known as **Softmax**. **Softmax** is very different from the other activation functions we have looked at so far because it computes a probability distribution for a list of values that are proportional to the relative scale of each of the values in the vector, the sum of which always equals 1.

Commonly used for **multiclass** classification, **Softmax** activation functions are used to normalize the outputs of a node, thus converting them from a weighted sum of values into **probabilities**, with its probability being a member of a given class. We can demonstrate this in Python using the numpy library:

```
def softmax_function(x):
    ex = np.exp(x - np.max(x))
    return ex / ex.sum()
```

```
x4 = [1, 2, 3, 4, 5]
y4 = softmax_function(x4)
print(y4)
```

Upon printing the values, we retrieve the following results:

```
[42]:   array([0.01165623, 0.03168492, 0.08612854, 0.23412166, 0.63640865])
```

Figure 8.14 – Results of a Softmax function

The two main advantages that come from using **Softmax** as an activation function are that the output values range between 0 and 1 and that they always sum to a value of 1.0. In return, this allows the function to be used to understand cross-entropy when it comes to the idea of divergence. We will visit this topic in more detail later in this chapter.

The various activation functions we have visited so far each have their pros and cons when it comes to using them in various applications. For example, sigmoid functions are commonly used for binary and multilabel classification applications, whereas **Softmax** functions are generally used for multiclass classification. This is not a hard rule, but simply a guide to help you match a function with the highest chance of success with its respective application:

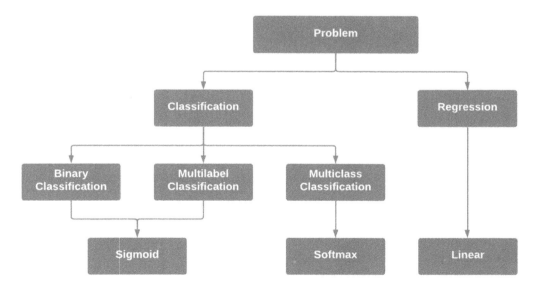

Figure 8.15 – Activation functions by problem type

Activation functions are a vital part of any deep learning model and are often regarded as *game changers* since simply changing one function for another can boost the performance of a model quite drastically. We will take a closer look at how model performance can be quantified within the scope of deep learning in the following section.

Measuring progress with loss

When we discussed the areas of classification and regression, we outlined a few measures to measure and quantify the performance of our models relative to one another. When it came to classification, we used **precision** and **accuracy**, whereas, in regression, we used **MAE** and **MSE**. Within the confines of deep learning, we will use a metric known as **loss**. The **loss** of a neural network is simply a measure of the cost that's incurred from making an incorrect prediction. Take, for example, a simple neural network with three input values and a single output value:

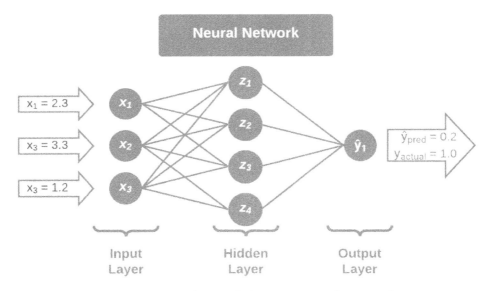

Figure 8.16 – A neural network showing input and output values

In this case, we have the values *[2.3, 3.3, 1.2]* being used as input values to the model, with a predicted value of 0.2 relative to the actual value of 1.0. We can demonstrate the loss as follows:

$$x^{(1)} = [2.3, 3.3, 1.2]$$

$$L(f(x^{(i)}; W), y)$$

In this function, $f(x^{(1)}; W)$ is the predicted value, while y is the actual value.

Empirical loss, on the other hand, is a measure of the total loss for the entirety of the dataset. We can represent empirical loss as follows:

$$J(W) = \frac{1}{n} \sum_{i=1}^{n} L(f(x^{(i)}; W), y^{(i)})$$

In this function, we sum the total losses for all calculations.

Throughout the model training process, our main objective will be to minimize this loss in a process known as **loss optimization**. The main idea behind loss optimization is to identify a set of weights that help achieve the lowest loss possible. We can visualize the idea of gradient descent as the process of moving from an initial starting value with a high loss, to a final value with a low loss. Our objective will be to ensure that converge in a global minimum rather than a local minimum, as depicted in the following diagram:

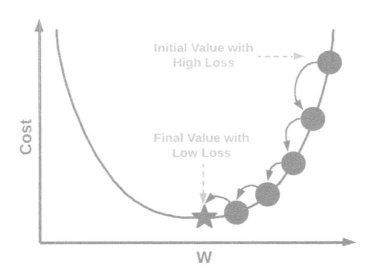

Figure 8.17 – The process of loss optimization

Each step we take to get closer to the minimum value is known as a **learning step**, the **learning rate** of which is generally determined by the user. This parameter is just one of many that we can specify using Keras, which we will learn about in the following section.

Deep learning with Keras

Within the realm of data science, the availability and use of various **frameworks** will always be crucial to standardizing the methods we use to develop and deploy models. So far, we have focused our machine learning efforts on using the scikit-learn framework. Throughout this section, we will learn about three new frameworks specifically focused on deep learning: **Keras**, **TensorFlow**, and **PyTorch**. These two frameworks are the two most popular amongst data scientists when it comes to developing various deep learning models as they offer a comprehensive list of APIs for numerous problems and use cases.

Understanding the differences between Keras and TensorFlow

Although these two platforms allow users to develop deep learning models, there are a few differences to know about. TensorFlow is known as an end-to-end machine learning platform that offers a comprehensive list of libraries, tools, and numerous resources. Users can manage data, develop models, and deploy solutions. Unlike most other libraries, **TensorFlow** offers both low and high levels of abstractions through their APIs, giving users lots of flexibility when it comes to developing models. On the other hand, **Keras** offers high-level APIs for developing neural networks, which run using **TensorFlow**. The high-level nature of this library allows users to begin developing and training complex neural networks with only a few lines of Python code. Keras is generally regarded as user-friendly, modular, and extendable. A third library exists that is commonly used in the deep learning space known as **PyTorch**. **PyTorch** is a low-level API known for its remarkable speed and optimization in the model training process. The architectures within this library are generally complex and not appropriate for introductory material, so they are not within the scope of this book. However, it is worth mentioning as it is one of the most common libraries in the machine learning space that you will likely encounter. Let's take a closer look at all three:

	KERAS	TENSORFLOW	PYTORCH
POPULARITY	1	2	3
API LEVEL	High-Level	Low and High-Level	Low-Level
ARCHITECTURE COMPLEXITY	Simple	Medium	Complex
DATASET SIZE	Small-Medium	Medium-Large	Medium-Large
KNOWN FOR	Backend Support	Object Detection	Fast Training

Figure 8.18 – A comparison between three of the most common deep learning frameworks

There are pros and cons for each of these libraries and you should select one of these libraries based on the task you set out to accomplish. Given that we are exploring the development of deep learning models for the first time, we will focus on using the Keras library.

Getting started with Keras and ANNs

Before we move on to a full tutorial, let's look at an example of using the Keras library since we have not explored its functionality and code:

1. First, we will need some sample data to use. Let's take advantage of the make_ blobs class in sklearn to create a classification dataset.

2. We will specify the need for two classes (binary classification) and a cluster standard deviation of 5 to ensure that the two clusters overlap, making it a more difficult dataset to work with:

```
from sklearn.datasets import make_blobs
X, y = make_blobs(n_samples=2000, centers=2, n_
features=4, random_state=1, cluster_std=5)
```

3. Next, we can scale the data using the MinMaxScaler() class:

```
from sklearn.preprocessing import MinMaxScaler
scalar = MinMaxScaler()
scalar.fit(X)
X_scaled = scalar.transform(X)
```

4. Following this transformation, we can split the data into training and testing sets, similar to how we have done previously:

```
from sklearn.model_selection import train_test_split
X_train, X_test, y_train, y_test = train_test_split(X_
scaled, y, test_size=0.25)
```

5. Let's go ahead and convert the array into a DataFrame to check the first few rows of data beforehand:

```
dfx_train = pd.DataFrame(X_train, columns=["Feature1",
"Feature2", "Feature3", "Feature4"])
dfx_train.head()
```

This will render the following table:

	Feature1	Feature2	Feature3	Feature4
0	0.433840	0.633580	0.647940	0.432928
1	0.614738	0.666253	0.408293	0.336253
2	0.345138	0.581183	0.307716	0.389190
3	0.746947	0.518434	0.541429	0.260596
4	0.410744	0.204487	0.438169	0.359704

Figure 8.19 – An example of the data within the DataFrame of features

6. We can check the overlap of the two clusters by using the `seaborn` library to plot the first two of the four features of the training dataset:

```
sns.scatterplot(x=dfx_train.Feature1, y=dfx_train.
Feature2, hue=y_train)
```

The following diagram shows the preceding code's output:

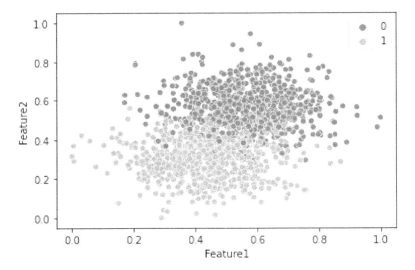

Figure 8.20 – A scatterplot of the dataset showing the overlapping nature of the two classes

Here, we can see that the data is quite well blended, making it difficult for some of the machine learning models we have explored to be able to separate the two classes with a high degree of **accuracy**.

7. With the data ready, we can go ahead and use the Keras library. One of the most popular methods for setting up a model is using the `Sequential()` class from `Keras`. Let's go ahead and import the class and instantiate a new model:

```
from keras.models import Sequential
model = Sequential()
```

8. With the model now instantiated, we can add a new layer to our model using the `Dense` class. We can also specify the number of nodes (4), `input_shape` (4 for the four features), `activation` (relu), and a unique name for the layer:

```
from keras.layers import Dense
model.add(Dense(4, input_shape=(4,), activation='relu',
name="DenseLayer1"))
```

9. To review the model we have built so far, we can use the `summary()` function:

```
model.summary()
```

This will give us some information and details about the model so far:

```
Model: "sequential_23"

_____
Layer (type)                 Output Shape              Param #
=================================================================
DenseLayer1 (Dense)          (None, 4)                 12
=================================================================
Total params: 12
Trainable params: 12
Non-trainable params: 0
```

Figure 8.21 – The sample output of the model's summary

10. We can add a few more layers to our model by simply using the `model.add()` function again after the fact, perhaps even with a different number of nodes:

```
model.add(Dense(8, activation='relu',
name="DenseLayer2"))
```

11. Since we are developing a **binary classification** model with two classes, 0 and 1, we can only have a single output value come from the model. Therefore, we will need to add one more layer that reduces the number of nodes from 8 to 1. In addition, we will change the activation to `sigmoid`:

```
model.add(Dense(1, activation='sigmoid',
name="DenseLayer3"))
```

12. Now that our model's general architecture has been set up, we will need to use the compile function and specify our loss. Since we are creating a binary classifier, we can use the `binary_crossentropy` loss and specify accuracy as our main metric of interest:

```
model.compile(loss='binary_crossentropy',
optimizer='adam', metrics=["accuracy"])
```

With the model ready, let's use the summary function to check it once more:

```
Model: "sequential_3"
_____
Layer (type)                 Output Shape              Param #
=================================================================
DenseLayer1 (Dense)          (None, 4)                 20
_____
DenseLayer2 (Dense)          (None, 8)                 40
_____
DenseLayer3 (Dense)          (None, 1)                 9
=================================================================
Total params: 69
Trainable params: 69
Non-trainable params: 0
```

Figure 8.22 – The sample output of the model's summary

So far, the model is quite simple. It will intake a dataset with four features in the first layer, expand that to eight nodes in the second layer, and then reduce it down to a single output in the third layer. With the model all set, we can go ahead and train it.

We can train a model by using the `model.fit()` function and by specifying the `X_train` and `y_train` sets. In addition, we will specify 50 `epochs` to train over. **Epochs** are simply the number of passes or iterations. We can also control the verbosity of the model, allowing us to control the amount of output data we want to see in the training process.

13. Recall that, in our earlier machine learning models, we only used the training data to train the model and kept the testing data to test the model after the training was completed. We will use the same methodology here as well; however, we will take advantage of the high-level nature of **Keras** and specify a `validation split` to be used in the training process. Deep learning models will almost always overfit your data. Using a validation split in the training process can help mitigate this:

```
history = model.fit(X_train, y_train, epochs=50,
    verbose=1, validation_split=0.2)
```

As the model begins the training process, it will begin to produce the following output. You can monitor the performance here by looking at the number of **epochs** on the left and the **metrics** on the right. When training a model, our objective is to ensure that the **loss** metric is constantly decreasing, whereas the accuracy is increasing. We can see an example of this in the following screenshot:

```
Epoch 1/75
38/38 [==============================] - 1s 10ms/step - loss: 0.6895 - accuracy: 0.5758 - val_loss: 0.6872 - val_accuracy: 0.6200
Epoch 2/75
38/38 [==============================] - 0s 2ms/step - loss: 0.6783 - accuracy: 0.7500 - val_loss: 0.6707 - val_accuracy: 0.8200
Epoch 3/75
38/38 [==============================] - 0s 2ms/step - loss: 0.6603 - accuracy: 0.8425 - val_loss: 0.6548 - val_accuracy: 0.8133
Epoch 4/75
38/38 [==============================] - 0s 2ms/step - loss: 0.6421 - accuracy: 0.8825 - val_loss: 0.6376 - val_accuracy: 0.8500
Epoch 5/75
38/38 [==============================] - 0s 2ms/step - loss: 0.6216 - accuracy: 0.9017 - val_loss: 0.6179 - val_accuracy: 0.8467
Epoch 6/75
38/38 [==============================] - 0s 2ms/step - loss: 0.5986 - accuracy: 0.9033 - val_loss: 0.5955 - val_accuracy: 0.8800
```

Figure 8.23 – A sample of a model's output

14. With the model trained, let's quickly examine the classification metrics, as we did previously, to get a sense of the performance. We can begin by making predictions using the testing data and using the `classification_report` to calculate our metric. Note that the `predict()` method does not return a class but a probability that needs to be rounded to either 0 or 1, given that this is a **binary classification** problem:

```
y_pred = (model.predict(X_test) > 0.5).astype("int32").
ravel()
from sklearn.metrics import classification_report
print(classification_report(y_pred, y_test))
```

Upon printing the report, we will get the following results:

	precision	recall	f1-score	support
0	0.95	0.93	0.94	266
1	0.92	0.94	0.93	234
accuracy			0.94	500
macro avg	0.94	0.94	0.94	500
weighted avg	0.94	0.94	0.94	500

Figure 8.24 – The results of the model

15. We can see that the **precision** and **recall** are quite high – not too bad for our first attempt! We can take a look at the model's performance from a different perspective using the `history` variable, which contains the model's training history:

```
fig = plt.figure(figsize=(10,10))

# total_rows, total_columns, subplot_index(1st, 2nd,
etc..)
plt.subplot(2, 2, 1)
plt.title("Accuracy", fontsize=15)
plt.xlabel("Epochs", fontsize=15)
plt.ylabel("Accuracy (%)", fontsize=15)
plt.plot(history.history["val_accuracy"],
label='Validation Accuracy', linestyle='dashed')
plt.plot(history.history["accuracy"], label='Training
Accuracy')
plt.legend(["Validation", "Training"], loc="lower right")

plt.subplot(2, 2, 2)
plt.title("Loss", fontsize=15)
plt.xlabel("Epochs", fontsize=15)
plt.ylabel("Loss", fontsize=15)
plt.plot(history.history["val_loss"], label='Validation
loss', linestyle='dashed')
plt.plot(history.history["loss"], label='Training loss')
plt.legend(["Validation", "Training"], loc="lower left")
```

Upon executing this code, we will receive the following diagram, which shows the change in accuracy and loss over the course of the model training process:

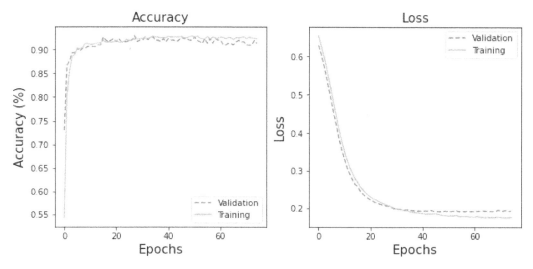

Figure 8.25 – The accuracy and loss of the model

When training a model, recall that the main objective is to ensure that the loss is decreasing, and never increasing over time. In addition, our secondary objective is to ensure that the accuracy of our model slowly and steadily increases. When trying to diagnose a model that is not performing well, the first step is to generate graphs such as these to get a sense of any potential problems before altering the model in an attempt to improve the metrics.

When we worked with most of the machine learning models in the previous chapters, we learned that we could alter these metrics by doing the following:

- Improving our data preprocessing.

- Tuning our hyperparameters or changing the model. Within the confines of deep learning, in addition to the options mentioned previously, there are a few more tools we have to change to suit our needs. For instance, we can change the overall architecture by adding or removing layers and nodes. In addition, we can change the activation functions within each of the layers to whatever would complement our problem statement the most. We can also change the optimizer or the learning rate of our optimizer (Adam, in this model).

With so many changes that can be made that could have high impacts on any given model, we will need to organize our work. We could either create numerous models and record our metrics manually in a spreadsheet, or we could take advantage of a library specifically designed to handle use cases such as these: **MLflow**. We will take a closer look at **MLflow** in the next section.

Tutorial – protein sequence classification via LSTMs using Keras and MLflow

Deep learning has gained a surge of popularity in recent years, prompting many scientists to turn to the field as a new means for solving and optimizing scientific problems. One of the most popular applications for deep learning within the biotechnology space involves **protein sequence** data. So far within this book, we have focused our efforts on developing predictive models when it comes to **structured** data. We will now turn our attention to data that's **sequential** in the sense that the elements within a sequence bear some relation to their previous element. Within this tutorial, we will attempt to develop a protein **sequence classification** model in which we will classify protein sequences based on their known family accession using the **Pfam** (`https://pfam.xfam.org/`) dataset.

> **Important note**
>
> `Pfam` dataset: Pfam: The protein families database in 2021 J. Mistry, S. Chuguransky, L. Williams, M. Qureshi, G.A. Salazar, E.L.L. Sonnhammer, S.C.E. Tosatto, L. Paladin, S. Raj, L.J. Richardson, R.D. Finn, A. BatemanNucleic Acids Research (2020) doi: 10.1093/nar/gkaa913 (`Pfam: The protein families database in 2021`).

The `Pfam` dataset consists of several columns, as follows:

- `Family_id`: The name of the family that the sequence belongs to (for example, filamin)

- `Family Accession`: The class or output that our model will aim to predict

- `Sequence`: The amino acid sequence we will use as input for our model

Throughout this tutorial, we will use the sequence data to develop several predictive models to determine each sequence's associated family accession. The sequences are in their raw state with different lengths and sizes. We will need to pre-process the data and structure it in such a way as to prepare it for sequence classification. When it comes to the labels, we will develop a model using a **balanced** set of different labels to ensure the model does not learn any particular bias.

As we begin to develop our ideal classification model, we will need to alter the many possible parameters to maximize the performance. To keep track of these changes, we will make use of the MLflow (`https://mlflow.org`) library. There are four main components within **MLflow**:

- **MLflow Tracking**: Allows users to record and query experiments
- **MLflow Projects**: Packages data science code
- **MLflow Models**: Deploys trained machine learning models
- **MLflow Registry**: Stores and manages your models

Within this tutorial, we will explore how to use MLflow tracking to track and manage the development of a protein sequence classification model. With these items in mind, let's get started.

Importing the necessary libraries and datasets

We will begin with the standard set of library imports, followed by the dataset in the format of a CSV document:

```
import pandas as pd
import numpy as np
from tensorflow.keras.utils import to_categorical
import matplotlib.pyplot as plt
import seaborn as sns
sns.set_style("darkgrid")
```

With the libraries now imported, we can import the dataset as well. We will begin by specifying the path, and then concatenate the dataset using a `for` loop:

```
PATH = "../../../datasets/dataset_pfam/"
files = []

for i in range(8):
    df = pd.read_csv(PATH+f"dataset_pfam_seq_sd{i+1}.csv",
index_col=None, header=0)
    files.append(df)
```

```
df = pd.concat(files, axis=0, ignore_index=True)
df.shape
```

Upon importing the dataset, we immediately notice that it contains five columns and ~1.3 million rows of data – slightly larger than what we have worked with so far. We can take a quick glimpse at the dataset using the `.head()` function:

	family_id	sequence_name	family_accession	aligned_sequence	sequence
0	GMC_oxred_C	A4WZS5_RHOS5/416-539	PF05199.13	PHPE.SRIRLST.RRDAHGMP.....IP.RIESRLGP..........	PHPESRIRLSTRRDAHGMPIPRIESRLGPDAFARLRFMARTCRAIL...
1	DUF2887	K9QI92_9NOSO/3-203	PF11103.8	RDSIYYQIFKRFPALIFEL..VD.NRPPQAQNYRFESVEVKETAFR...	RDSIYYQIFKRFPALIFELVDNRPPQAQNYRFESVEVKETAFRIDG...
2	zf-IS66	Q92LC9_RHIME/32-75	PF13005.7	.TCC.PDCGG.E..LRLVGED.AS....EILDMI.AAQMKVIEVARL...	TCCPDCGGELRLVGEDASEILDMIAAQMKVIEVARLKKSCRCCE
3	Asp_decarbox	X2GQZ4_9BACI/1-115	PF02261.16	MLRMMMNSKIHRATVTEADLNYVGSITIDEDILDAVGMLPNEKVHI...	MLRMMMNSKIHRATVTEADLNYVGSITIDEDILDAVGMLPNEKVHI...
4	Filamin	A7SQM3_NEMVE/342-439	PF00630.19	TACPKQ.CTA....RGLG...........LK.AAPVT.QPT..R...	TACPKQCTARGLGLKAAPVTQPTRFVVILNDCHGQPLGRSEGELEV...

Figure 8.26 – A sample of the data from the protein sequence dataset

With the dataset successfully imported, let's go ahead and explore the dataset in more detail.

Checking the dataset

We can confirm the completeness of the data in this DataFrame using the `isna()` function, followed by the `sum()` function to summarize by column:

```
df.isna().sum()
```

Now, let's take a closer look at the `family_accession` column (our model's output) of this dataset. We can check the total number of instances by grouping the column and using the `value_counts()` function, followed by the `n_largest()` function, to get the top 10 most common entries in this column:

```
df["family_accession"].groupby(df["family_accession"]).value_
counts().nlargest(10)
```

Grouping the data will yield the following results:

```
family_accession   family_accession
PF13649.6          PF13649.6                 4545
PF00560.33         PF00560.33                2407
PF13508.7          PF13508.7                 2199
PF06580.13         PF06580.13                1921
PF02397.16         PF02397.16                1908
PF00677.17         PF00677.17                1878
PF01035.20         PF01035.20                1681
PF02417.15         PF02417.15                1579
PF13472.6          PF13472.6                 1564
PF00684.19         PF00684.19                1512
Name: family_accession, dtype: int64
```

Figure 8.27 – A summary of the classes in the dataset with value counts higher than 1,200

Here, we can see that 1,500 entries seems to be the cutoff point for the top 10 values. We can also take a closer look at the sequence column (our model's input) by getting a sense of the average lengths of the sequences. We can plot the count of each sequence length using the `displot()` function from the `seaborn` library:

```
sns.displot(df["sequence"].apply(lambda x: len(x)), bins=75,
height=4, aspect=2)
```

Executing this code will yield the following results:

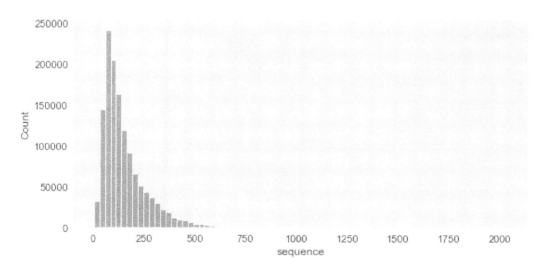

Figure 8.28 – A histogram of the counts of the sequence lengths in the dataset

From this graph, as well as by using the `mean()` and `median()` functions, we can see that the average and most common lengths are approximately 155 and 100 amino acids. We will use these numbers later when determining what the cutoff should be for the input sequences.

Now that we have gained a better sense of the data, it is time to prepare the dataset for our classification models. We could theoretically train the model on the dataset as a whole without limits – however, the models would require a much longer duration to train. In addition, by training across all the data without accounting for balance, we may introduce bias within the model. To mitigate both of these situations, let's reduce this dataset by filtering for the classifications with at least 1,200 **observations**:

```
df_filt = df.groupby("family_accession").filter(lambda x:
len(x) > 1200)
```

Given that some classes have significantly more than 1,200 observations, we can randomly select exactly 1,200 observations using the `sample()` function:

```
df_bal = df_filt.groupby('family_accession').apply(lambda x:
x.sample(1200))
```

We can now check the filtered and balanced dataset using the `head()` function:

```
df_red = df_bal[["family_accession", "sequence"]].reset_
index(drop=True)
df_red.head()
```

The `head()` function will yield the following results:

	family_accession	sequence
0	PF00126.27	YKHLVLIDTLARTRNMHATATRMNLSQPALSKMLRDLEEQFGFALF...
1	PF00126.27	LRDLRHFLAVAEEGHIGRAAARLHLSQPPLTRHIQALEEKIGVPLF...
2	PF00126.27	RQHLSILREVDRMGSLTAAAERLNVSQSALSHTIRKLEDRYGVAMW...
3	PF00126.27	LNLVVALRALLEERNVTRAGQRVGLSQPAMSAALARLRRHFDDDLL...
4	PF00126.27	LKRIVIFNKVVECGSFTCAAEALGMTKSKVSEQITALEKTLNVRLL...

Figure 8.29 – A sample of the data in the form of a DataFrame

We can check the number of classes we will have in this dataset by checking the length of the `value_counts()` function:

```
num_classes = len(df_red.family_accession.value_counts())
```

If we check the `num_classes` variable, we will see that we have 28 possible classes in total.

Splitting the dataset

With the data prepared, our next step will be to split the dataset into training, testing, and validation sets. We will once again make use of the `train_test_split` function from `sklearn` to accomplish this:

```
from sklearn.model_selection import train_test_split
X_train, X_test = train_test_split(df_red, test_size=0.25)
X_val, X_test = train_test_split(X_test, test_size=0.50)
```

With the data now split, let's go ahead and preprocess it.

Preprocessing the data

With the data split up, we need to preprocess the datasets to use on our neural network models. First, we will need to reduce the sequences down to the 20 most common amino acids and convert the sequences into integers. This will speed up the training process. First, we will create a dictionary of amino acids that contains their corresponding values:

```
aa_seq_dict = {'A': 1,'C': 2,'D': 3,'E': 4,'F': 5,'G': 6,'H': 7,'I': 8,'K': 9,'L': 10,'M': 11,'N': 12,'P': 13,'Q': 14,'R': 15,'S': 16,'T': 17,'V': 18,'W': 19,'Y': 20}
```

Next, we can iterate over the sequences and convert the string values into their corresponding integers. Note that we will complete this for the training, testing, and validation sets:

```
def aa_seq_encoder(data):
    full_sequence_list = []
    for i in data['sequence'].values:
        row_sequence_list = []
        for j in i:
            row_sequence_list.append(aa_seq_dict.get(j, 0))
        full_sequence_list.append(np.array(row_sequence_list))
```

```
        return full_sequence_list

X_train_encode = aa_seq_encoder(X_train)
X_val_encode = aa_seq_encoder(X_val)
X_test_encode = aa_seq_encoder(X_test)
```

Next, we will need to pad the sequences to ensure they are all of an equal length. To accomplish this, we can use the `pad_sequences` function from `keras`. We will specify `max_length` for each of the sequences as 100, given that it approximates the median value we saw earlier. In addition, we will pad the sequences with `'post'` to ensure that we pad them at the end instead of the front:

```
from keras.preprocessing.sequence import pad_sequences
max_length = 100
X_train_padded = pad_sequences(X_train_encode, maxlen=max_
length, padding='post', truncating='post')
X_val_padded = pad_sequences(X_val_encode, maxlen=max_length,
padding='post', truncating='post')
X_test_padded = pad_sequences(X_test_encode, maxlen=max_length,
padding='post', truncating='post')
```

We can take a quick glance of the changes we have made using one of the sequences. First, we have the raw sequence as a `string`:

```
X_train.sequence[1][:30]
'LRDLRHFLAVAEEGHIGRAAARLHLSQPPL'
```

Next, we can encode the sequence to remove uncommon amino acids and convert the string to a list of integers:

```
X_train_encode[1][:30]
array([ 7, 10, 15, 18, 10,  3, 18, 16, 14, 17, 15,  5, 12,
       10,  7, 16, 15, 12, 12,  8, 18,  4, 14,  5, 17,  4,  2])
```

Finally, we can limit the lengths of the sequences to either truncate them at 100 elements or **pad** them with zeros to reach 100 elements:

```
X_train_padded[1][:30]
array([ 7, 10, 15, 18, 10,  3, 18, 16, 14, 17, 15,  5,
       12, 10,  7, 16, 15, 12, 12,  8, 18,  4, 14,  5,
       17,  4,  2,  0,  0,  0])
```

Now that we have preprocessed the input data, we will need to preprocess the output values as well. We can do so using the `LabelEncoder` class from `sklearn`. Our main objective here will be to transform the values from a list of labels in a dataframe column into an encoded list:

```
from sklearn.preprocessing import LabelEncoder
le = LabelEncoder()
y_train_enc = le.fit_transform(X_train['family_accession'])
y_val_enc = le.transform(X_val['family_accession'])
y_test_enc = le.transform(X_test['family_accession'])
```

Finally, we can use the `to_categorical` function from `sklearn` to transform a class vector into a binary class matrix:

```
from tensorflow.keras.utils import to_categorical
y_train = to_categorical(y_train_enc)
y_val = to_categorical(y_val_enc)
y_test = to_categorical(y_test_enc)
```

To review the changes we've made here, we can use a single column in a `DataFrame`:

```
X_train['family_accession']
```

We can see the results of this in the following diagram:

```
5190      PF00586.24
3971      PF00560.33
25388     PF03453.17
14023     PF01523.16
12758     PF01368.20
             . . .
24485     PF02885.17
11655     PF01255.19
13012     PF01368.20
18836     PF02397.16
15080     PF01715.17
Name: family_accession, Length: 25200, dtype: object
```

Figure 8.30 – A list of the classes

Next, we must encode the classes into a list of numerical values, with each value representing a specific class:

```
y_train_enc
array([ 4,   3, 21, ...,  10, 15, 12], dtype=int64)
```

Finally, we must convert the structure into a **binary class matrix** so that each row consists of a list of 27 values of zero, and one value of 1, representing the class it belongs to:

```
y_train[5]
array([0., 0., 0., 0., 0., 0., 0., 0., 0., 0., 0., 0., 0.,
0., 0., 0., 0., 1., 0., 0., 0., 0., 0., 0., 0., 0., 0.],
dtype=float32)
```

With that, our datasets have been fully preprocessed and ready to be used in the model development phase of the tutorial.

Developing models with Keras and MLflow

Similar to our previous example of developing models with the `Keras` library, we will once again be using the `Sequential` class:

1. We will begin by importing the layers and other items we will need from Keras:

    ```
    import tensorflow as tf
    from keras.models import Sequential
    from keras.layers import Dense, Conv1D, MaxPooling1D,
    Flatten, Input, Bidirectional, LSTM, Dropout
    from keras.layers.embeddings import Embedding
    from keras.regularizers import l2
    from keras.models import Model


    import mlflow
    import mlflow.keras
    ```

2. Now, we can create a new instance of a model using the sequential class and begin populating it by adding a few layers of interest. We will start by adding an embedding layer to convert positive integers into dense vectors. We will specify an `input_dim` of 21 to represent the size of the amino acid index + 1, and an `output_dim` of 32. In addition, we will assign an `input_length` equal to that of `max_length` – the length of the sequences:

```
model = Sequential()
model.add(Embedding(21, 8, input_length=max_length,
name="EmbeddingLayer"))
```

3. Next, we will add an LSTM layer, wrapped in a `Bidirectional` layer, to run inputs in both directions – from past to future and from future to past:

```
model.add(Bidirectional(LSTM(8),
name="BidirectionalLayer"))
```

4. Next, we will add a `Dropout` layer to help prevent the model from overfitting:

```
model.add(Dropout(0.2, name="DropoutLayer"))
```

5. Finally, we will end with a `Dense` layer and set the number of nodes to 28 so that this corresponds with the shape of the outputs. Notice that we use a Softmax activation here:

```
model.add(Dense(28, activation='softmax',
name="DenseLayer"))
```

6. With the model's architecture prepared, we can assign an optimizer (Adam), compile the model, and check the summary:

```
opt = tf.keras.optimizers.Adam(learning_rate=0.1)
model.compile(optimizer=opt, loss='categorical_
crossentropy', metrics=['accuracy'])
```

7. Now, let's go ahead and train our model using the `fit()` function and assign 30 epochs. Notice from our previous tutorial that training deep learning models can be very time-consuming and expensive, so training a model that is not learning can be a major waste of time. To mitigate situations such as these, we can implement what is known as a callback in the sense that Keras can end the training period when a model is no longer learning (that is, the loss is no longer decreasing):

```
from keras.callbacks import EarlyStopping
es = EarlyStopping(monitor='val_loss', patience=5,
verbose=1)
```

8. Finally, we can go ahead and log our new run in MLflow by calling the `autolog()` function and fitting the model, as we did previously. MLflow offers many different methods to log both parameters and metrics, and you are not limited to using just `autolog()`:

```
mlflow.keras.autolog()
history = model.fit(
    X_train_padded, y_train,
    epochs=30, batch_size=256,
    validation_data=(X_val_padded, y_val),
    callbacks=[es]
    )
```

Assuming you followed these steps correctly, the model will print a note stating that MLflow is being used, and you should see a new directory appear next to your current notebook. Upon completing the training process, you can plot the results, as we did previously, to arrive at the following diagram:

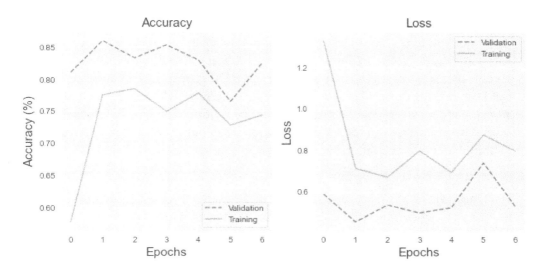

Figure 8.31 – The accuracy and results of the first iteration of this model

Here, we can see that the accuracy seems to remain stagnant at around 80-85%, while the loss remains stagnant at 0.6 to 0.8. We can see that the model is not learning. Perhaps a change of parameters is needed? Let's go ahead and change the number of nodes from 8 to 12 and the learning rate from 0.1 to 0.01. Upon compiling the new model, calling the autolog() function, and training the new dataset, we will arrive at a new diagram:

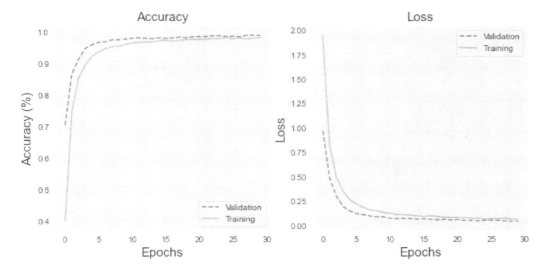

Figure 8.32 – The accuracy and results of the next iteration of this model

Here, we can see that the model's loss, both for training and validation, decreased quite nicely until the callback stopped the training at around 30 epochs in. Alternatively, the accuracy shows a sharp increase at the beginning, followed by a stable increase toward the end, also stopping at 30 epochs into the process. We can keep making our changes and calling the `autolog()` function over and over, allowing the system to log the changes and the resulting metrics on our behalf. After several iterations, we can review the performance of our models using `mlflow ui`. Within the notebook itself, enter the following command:

```
!mlflow ui
```

Next, navigate to `http://localhost:5000/`. There, you will be able to see the MLflow UI, where you will be able to view the models, their parameters, and their associated metrics:

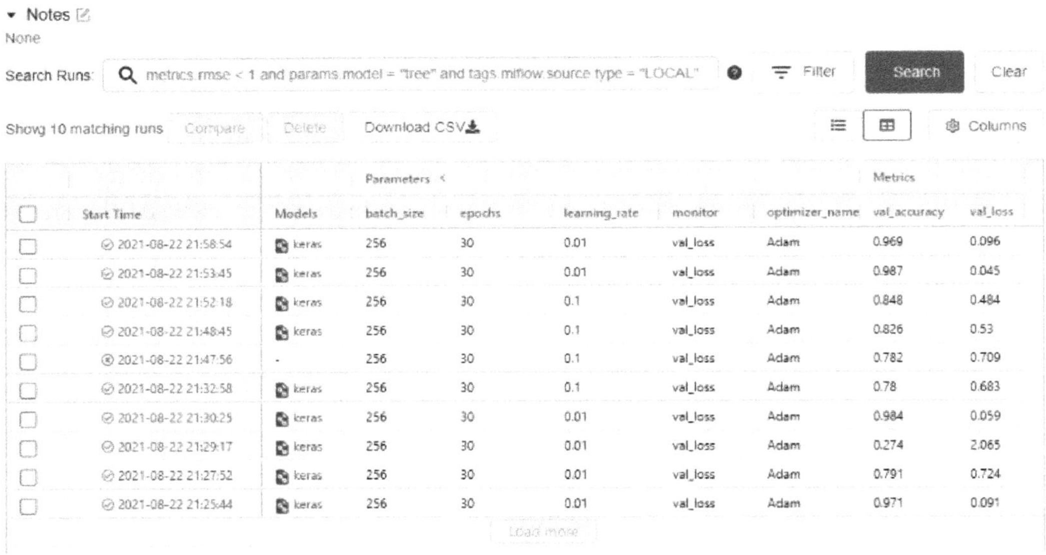

▼ Notes ✎
None

Search Runs: 🔍 metrics.rmse < 1 and params.model = "tree" and tags.mlflow.source.type = "LOCAL"   ❷   ☰ Filter   [Search]   Clear

Showg 10 matching runs   Compare   Delete   Download CSV⬇   ≡   ▦   ⚙ Columns

	Start Time	Models	batch_size	epochs	learning_rate	monitor	optimizer_name	val_accuracy	val_loss
☐	⊘ 2021-08-22 21:58:54	keras	256	30	0.01	val_loss	Adam	0.969	0.096
☐	⊘ 2021-08-22 21:53:45	keras	256	30	0.01	val_loss	Adam	0.987	0.045
☐	⊘ 2021-08-22 21:52:18	keras	256	30	0.1	val_loss	Adam	0.848	0.484
☐	⊘ 2021-08-22 21:48:45	keras	256	30	0.1	val_loss	Adam	0.826	0.53
☐	⊗ 2021-08-22 21:47:56	-	256	30	0.1	val_loss	Adam	0.782	0.709
☐	⊘ 2021-08-22 21:32:58	keras	256	30	0.1	val_loss	Adam	0.78	0.683
☐	⊘ 2021-08-22 21:30:25	keras	256	30	0.01	val_loss	Adam	0.984	0.059
☐	⊘ 2021-08-22 21:29:17	keras	256	30	0.01	val_loss	Adam	0.274	2.065
☐	⊘ 2021-08-22 21:27:52	keras	256	30	0.01	val_loss	Adam	0.791	0.724
☐	⊘ 2021-08-22 21:25:44	keras	256	30	0.01	val_loss	Adam	0.971	0.091

Load more

Figure 8.33 – An example of the MLflow UI

With that, you can select the best model and move forward with your project.

Reviewing the model's performance

Now that the best-performing model has been selected, let's get a better sense of its associated **metrics**. We can gain a better sense of the results by using a classification_report, as we did previously, showing almost 99% for both precision and recall. Alternatively, we can use a confusion matrix to get a better sense of the data, given that we have 28 classes in total:

```
from sklearn.metrics import confusion_matrix
y_pred = model.predict(X_test_padded)
cf_matrix = confusion_matrix(np.argmax(y_test, axis=1),
np.argmax(y_pred, axis=1))
```

With the confusion matrix calculated, we can use a heatmap to visualize the results:

```
import seaborn as sns
plt.figure(figsize=(15,10))
sns.heatmap(cf_matrix, annot=True, fmt='', cmap='Blues')
```

Upon executing this, we will get the following diagram:

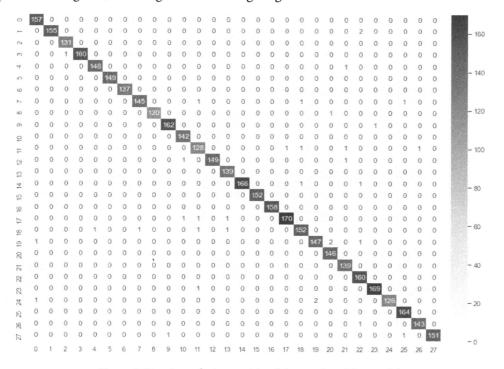

Figure 8.34 – A confusion matrix of the results of the model

Here, we can see that the performance of the model is quite robust as it is giving us great results! Keras, TensorFlow, and PyTorch are great packages that can help us develop robust and high-impact models to solve specific solutions. Often, we will find that there may be a model (or set of models) that already exists through AWS that can solve our complex problem with little to no code. We will explore an example of this in the next section.

Tutorial – anomaly detection in manufacturing using AWS Lookout for Vision

In the previous section, we prepared and trained a deep learning model to classify proteins in their given categories. We went through the process of preprocessing our data, developing a model, testing the parameters, editing the architecture, and selecting a combination that maximized our metrics of interest. While this process can generally produce good results, we can sometimes utilize platform architectures such as those from AWS to automatically develop models on our behalf. Within this tutorial, we will take advantage of a tool known **AWS Lookout for Vision** (`https://aws.amazon.com/lookout-for-vision/`) to help us prepare a model capable of detecting anomalies within a dataset.

Throughout this tutorial, we will be working with a dataset consisting of images concerned with manufacturing a of **Drug Product** (**DP**). Each of the images consists of a vial whose image was captured at the end of the manufacturing cycle. Most of the vials are clean and don't have any impurities. However, some of the vials contain minor impurities, as illustrated in the following diagram:

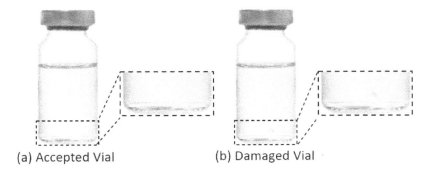

(a) Accepted Vial   (b) Damaged Vial

Figure 8.35 – An example of an accepted vial versus a damaged vial

The process of rejecting damaged or impure vials is often done manually and can be quite time-consuming. We have been tasked with implementing an automated solution to this problem and we only have a few days to do so. Rather than developing our own custom deep learning model for detecting anomalies in images, we can utilize **Amazon Lookout for Vision**. In this tutorial, we will begin by uploading our dataset of images to S3, importing the images into the framework, and begin training our model. With that in mind, let's go ahead and get started!

Within this book's GitHub repository, you can find a directory called `vials_input_dataset_s3`, which contains a collection of both normal and damaged vials. If we take a closer look at our dataset, we will notice that it is constructed using a directory hierarchy, as shown in the following diagram:

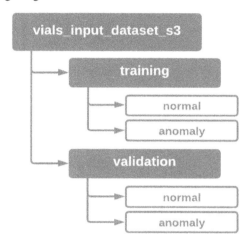

Figure 8.36 – An example of an accepted vial versus a damaged vial

We will begin by importing the images into the same S3 bucket we have been working with throughout this book:

1. First, navigate to S3 from within the AWS console and select the bucket of interest. In this case, I will select **biotech-machine-learning**.

2. Next, click the orange **Upload** button, select the **vials_input_dataset_s3** folder, and click **Upload**. This process may take a few moments, depending on your internet connection.

3. Now, click on the **Copy S3 URI** button at the top right-hand side of the page. We will need this URI in a few moments.

Now, our data is available for use to use in our S3 bucket. Next, we can focus on getting the data imported with the model and start the model training process:

1. To begin, navigate to Amazon Lookout for Vision, which is located in the AWS console. Then, click the **Get started** button:

Figure 8.37 – The front page of Amazon Lookout for Vision

2. Click on the **Create project** button on the right-hand side of the page and give your project a name.

3. Once the project has been created, go ahead and click the **Create dataset** button on the left-hand side of the page.

4. Select the second option to **Create a training dataset and test dataset**:

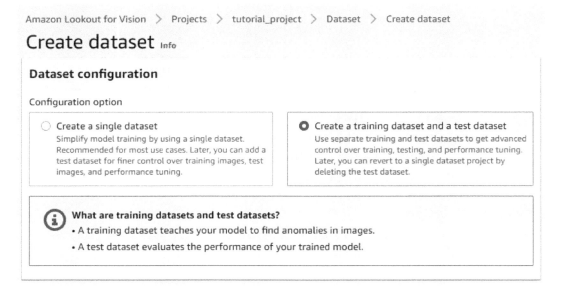

Figure 8.38 – Creating a dataset in AWS Lookout for Vision

5. Next, within the **Training dataset details** section, select the option to **Import images from S3 bucket** and paste the URI you copied previously into the S3 URI field. Since this section pertains to the training dataset, we will add the word `training` to the path, as shown in the following screenshot:

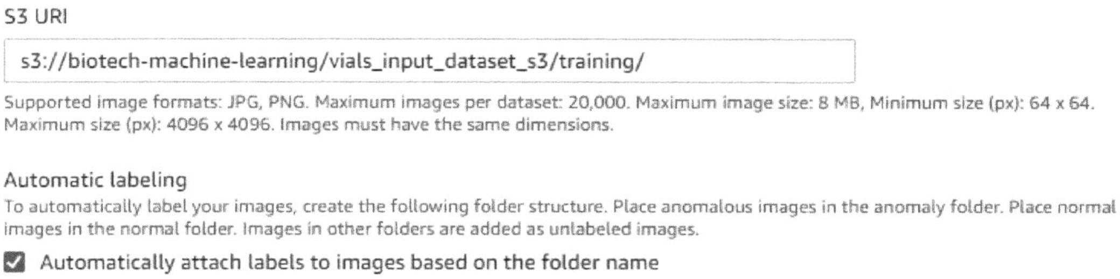

S3 URI
s3://biotech-machine-learning/vials_input_dataset_s3/training/
Supported image formats: JPG, PNG. Maximum images per dataset: 20,000. Maximum image size: 8 MB, Minimum size (px): 64 x 64. Maximum size (px): 4096 x 4096. Images must have the same dimensions.
Automatic labeling
To automatically label your images, create the following folder structure. Place anomalous images in the anomaly folder. Place normal images in the normal folder. Images in other folders are added as unlabeled images.
Automatically attach labels to images based on the folder name

Figure 8.39 – Creating a dataset in AWS Lookout for Vision

6. In addition, be sure to select the **Automatic labeling** option to ensure our labels are taken in by AWS.
7. Repeat this same process for the test dataset but be sure to add the word `validation` instead of training in the S3 URI path. Then, click on **Create dataset**.
8. Once the dataset has been created, you will be taken to a new page where you can visually inspect the dataset's readiness. Then, you can click the **Train model** button located in the top right-hand corner of the page to begin the model training process. This process can be time-consuming and may take a few hours:

vial_damaged_500005.png
vial_damaged_500487.png
vial_damaged_501421.png
Anomaly
Anomaly
Anomaly

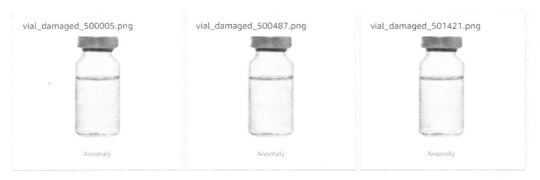

Figure 8.40 – The dataset before training the model

By doing this, you will be presented with the final results for the model, which will show you the precision, recall, and F1 score, as shown in the following screenshot:

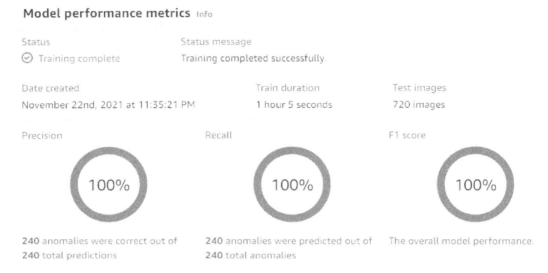

Figure 8.41 – The Model performance metrics page

With that final step completed, we have successfully developed a robust model capable of detecting anomalies in the manufacturing process! Not only were we able to create the models in just a few hours, but we managed to do so without any code!

Summary

Throughout this chapter, we made a major stride to cover a respectable portion of the *must-know* elements of deep learning and neural networks. First, we investigated the roots of neural networks and how they came about and then dove into the idea of a perceptron and its basic form of functionality. We then embarked on a journey to explore four of the most common neural networks out there: MLP, CNN, RNN, and LSTM. We gained a better sense of how to select activation functions, measure loss, and implement our understandings using the Keras library.

Next, we took a less theoretical and much more hands-on approach as we tackled our first dataset that was sequential nature. We spent a considerable amount of time preprocessing our data, developing our model, getting our model development organized with MLflow, and reviewing its performance. Following these steps allowed us to create a custom and well-suited model for the problem at hand. Finally, we took a no-code approach by using AWS Lookout for Vision to train a model capable of detecting anomalies in images of vials.

Deep learning and the application of neural networks have most certainly seen a major surge over the last few years, and in the next chapter, we will see an application of deep learning as it relates to natural language processing.

9
Natural Language Processing

In the previous chapter, we discussed using deep learning to not only address structured data in the form of tables but also sequence-based data where the order of the elements matters. In this chapter, we will be discussing another form of sequence-based data – text, within a field known as **Natural Language Processing** (**NLP**). We can define NLP as a subset of artificial intelligence that overlaps with both the realms of machine learning and deep learning, specifically when it comes to interactions between the areas of linguistics and computer science.

There are many well-known and well-documented applications and success stories of using NLP for various tasks. Products ranging from spam detectors all the way to document analyzers involve NLP to some extent. Throughout this chapter, we will explore several different areas and applications involving NLP.

As we have observed with many other areas of data science we have explored thus far, the field of NLP is just as vast and sparse, with endless tools and applications that a single book would never be able to fully cover. Throughout this chapter, we will aim to highlight as many of the most common and useful applications you will likely encounter as we can.

Throughout this chapter, we will explore many of the popular areas relating to NLP from the perspective of both structured and unstructured data. We will explore several topics, such as entity recognition, sentence analysis, topic modeling, sentiment analysis, and natural language search engines.

In this chapter, we will cover the following topics:

- Introduction to NLP

- Getting started with NLP using NLTK and SciPy

- Working with structured data

- Tutorial – abstract clustering and topic modeling

- Working with unstructured data

- Tutorial – developing a scientific data search engine using transformers

With these objectives in mind, let's go ahead and get started.

Introduction to NLP

Within the scope of **biotechnology**, we often turn to **NLP** for numerous reasons, which generally involve the need to organize data and develop models to find answers to scientific questions. As opposed to the many other areas we have investigated so far, NLP is unique in the sense that we focus on one type of data at hand: text data. When we think of text data within the realm of NLP, we can divide things into two general categories: **structured data** and **unstructured data**. We can think of structured data as text fields living within tables and databases in which items are organized, labeled, and linked together for easier retrieval, such as a SQL or DynamoDB database. On the other hand, we have what is known as unstructured data such as documents, PDFs, and images, which can contain static content that is neither searchable nor easily accessible. An example of this can be seen in the following diagram:

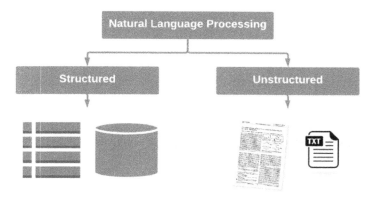

Figure 9.1 – Structured and unstructured data in NLP

Often, we wish to use documents or text-based data for various purposes, such as the following:

- **Generating insights**: Looking for trends, keywords, or key phrases

- **Classification**: Automatically labeling documents for various purposes

- **Clustering**: Grouping documents together based on features and characteristics

- **Searching**: Quickly finding important knowledge in historical documents

In each of these examples, we would need to take our data from unstructured data and move it toward a structured state to implement these tasks. In this chapter, we will look at some of the most important and useful concepts and tools you should know about concerning the NLP space.

Getting started with NLP using NLTK and SciPy

There are many different NLP libraries available in the Python language that allow users to accomplish a variety of different tasks for analyzing data, generating insights, or preparing predictive models. To begin our journey in the realm of NLP, we will take advantage of two popular libraries known as **NLTK** and **SciPy**. We will begin by importing these two libraries:

```
import nltk
import scipy
```

Often, we will want to parse and analyze raw pieces of text for particular purposes. Take, for example, the following paragraph regarding the field of biotechnology:

```
paragraph = """Biotechnology is a broad area of biology,
involving the use of living systems and organisms to develop
or make products. Depending on the tools and applications, it
often overlaps with related scientific fields. In the late
20th and early 21st centuries, biotechnology has expanded to
include new and diverse sciences, such as genomics, recombinant
gene techniques, applied immunology, and development of
pharmaceutical therapies and diagnostic tests. The term
biotechnology was first used by Karl Ereky in 1919, meaning
the production of products from raw materials with the aid of
living organisms."""
```

Here, there is a single string that's been assigned to the paragraph variable. Paragraphs can be separated into sentences using the sent_tokenize() function, as follows:

```
from nltk.tokenize import sent_tokenize
nltk.download('popular')
sentences = sent_tokenize(paragraph)
print(sentences)
```

Upon printing these sentences, we will get the following output:

```
['Biotechnology is a broad area of biology, involving the use of living systems and organisms to develop or make products.',
 'Depending on the tools and applications, it often overlaps with related scientific fields.',
 'In the late 20th and early 21st centuries, biotechnology has expanded to include new and diverse sciences, such as genomics,
recombinant gene techniques, applied immunology, and development of pharmaceutical therapies and diagnostic tests.',
 'The term biotechnology was first used by Karl Ereky in 1919, meaning the production of products from raw materials with the
aid of living organisms.']
```

Figure 9.2 – A sample list of sentences that were split from the initial paragraph

Similarly, we can use the word_tokenize() function to separate the paragraph into individual words:

```
from nltk.tokenize import word_tokenize
words = word_tokenize(sentences[0])
print(words)
```

Upon printing the results, you will get a list similar to the one shown in the following screenshot:

```
['Biotechnology',
 'is',
 'a',
 'broad',
 'area',
 'of',
 'biology',
 ',',
 'involving',
```

Figure 9.3 – A sample list of words that were split from the first sentence

Often, we will want to know the part of speech for a given word in a sentence. We can use the pos_tag() function on a given string to do this – in this case, the first sentence of the paragraph:

```
tokens = word_tokenize(sentences[0])
tags = nltk.pos_tag(tokens)
print(tags)
```

Upon printing the tags, we get a list of sets where the word or token is listed on the left and the associated part of speech is on the right. For example, Biotechnology and biology were tagged as proper nouns, whereas involving and develop were tagged as verbs. We can see an example of these results in the following screenshot:

```
[('Biotechnology', 'NNP'),
 ('is', 'VBZ'),
 ('a', 'DT'),
 ('broad', 'JJ'),
 ('area', 'NN'),
 ('of', 'IN'),
 ('biology', 'NN'),
 (',', ','),
 ('involving', 'VBG'),
 ('the', 'DT'),
 ('use', 'NN'),
 ('of', 'IN'),
 ('living', 'VBG'),
 ('systems', 'NNS'),
 ('and', 'CC'),
 ('organisms', 'NNS'),
 ('to', 'TO'),
 ('develop', 'VB'),
 ('or', 'CC'),
 ('make', 'VB'),
 ('products', 'NNS'),
 ('.', '.')]
```

Figure 9.4 – Results of the part-of-speech tagging method

In addition to understanding the parts of speech, we often want to know the frequency of the given words in a particular string of text. For this, we could either tokenize the paragraph into words, group by word, count the instances and plot them, or use NLTK's built-in functionality:

```
freqdist = nltk.FreqDist(word_tokenize(paragraph))
import matplotlib.pyplot as plt
import seaborn as sns
plt.figure(figsize=(10,3))
plt.xlabel("Samples", fontsize=20)
plt.xticks(fontsize=14)
plt.ylabel("Counts", fontsize=20)
plt.yticks(fontsize=14)
sns.set_style("darkgrid")
freqdist.plot(30,cumulative=False)
```

By doing this, you will receive the following output:

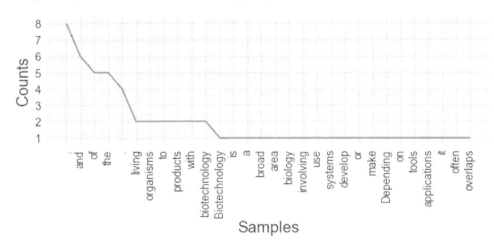

Figure 9.5 – Results of calculating the frequency (with stop words)

Here, we can see that the most common elements are commas, periods, and other irrelevant words. Words that are irrelevant to any given analysis are known as **stop words** and are generally removed as part of the preprocessing step. Punctuation, on the other hand, can be handled with regex. Let's go ahead and prepare a function to clean our text:

```
from nltk.corpus import stopwords
from nltk.tokenize import word_tokenize
```

```
nltk.download('punkt')
nltk.download('stopwords')
import re
STOP_WORDS = stopwords.words()


def cleaner(text):
    text = text.lower() #Convert to lower case
    text = re.sub("[^a-zA-Z]+", ' ', text) # Only keep text,
remove punctuation and numbers
    text_tokens = word_tokenize(text) #Tokenize the words
    tokens_without_sw = [word for word in text_tokens if not
word in STOP_WORDS] #Remove the stop words
    filtered_sentence = (" ").join(tokens_without_sw) # Join
all the words or tokens back to a single string
    return filtered_sentence
```

There are four main steps within this function. First, we convert the text into lowercase for consistency; then, we use regex to remove all punctuation and numbers. After, we split the string into individual tokens and remove the words if they are in our list of stop words, before finally joining the words back together into a single string.

> **Important note**
> Please note that text cleaning scripts are often specific to the use case in the sense that not all use cases require the same steps.

Now, we can apply the `cleaner` function to our paragraph:

```
clean_paragraph = cleaner(paragraph)
clean_paragraph
```

We can see the output of this function in the following screenshot:

```
'biotechnology broad area biology involving use living systems organisms develop make products depen
ding tools applications often overlaps related scientific fields late th early st centuries biotechn
ology expanded include new diverse sciences genomics recombinant gene techniques applied immunology
development pharmaceutical therapies diagnostic tests term biotechnology first used karl ereky meani
ng production products raw materials aid living organisms'
```

Figure 9.6 – Output of the text cleaning function

Upon recalculating the frequencies with the clean text, we can replot the data and view the results:

```
import matplotlib.pyplot as plt
import seaborn as sns
plt.figure(figsize=(10,3))
plt.xlabel("Samples", fontsize=20)
plt.xticks(fontsize=14)

plt.ylabel("Counts", fontsize=20)
plt.yticks(fontsize=14)

sns.set_style("darkgrid")
freqdist.plot(30,cumulative=False)
```

The output of this code can be seen in the following screenshot:

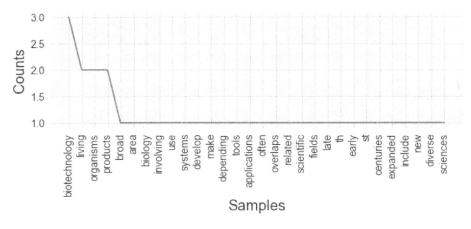

Figure 9.7 – Results of calculating the frequency (without stop words)

As we begin to dive deeper into our text, we will often want to tag items not only by their parts of speech, but also by their entities, allowing us to parse dates, names, and many others in a process known as **Named Entity Recognition** (**NER**). To accomplish this, we can make use of the spacy library:

```
import spacy
spacy_paragraph = nlp(paragraph)
spacy_paragraph = nlp(paragraph)
print([(X.text, X.label_) for X in spacy_paragraph.ents])
```

Upon printing the results, we obtain a list of items and their associated entity tags. Notice that the model not only picked up the year `1919` as a `DATE` entity but also picked up descriptions such as `21st centuries` as `DATE` entities:

```
[('the late 20th and early 21st centuries', 'DATE'), ('Karl Ereky', 'PERSON'), ('1919', 'DATE')]
```

Figure 9.8 – Results of the NER model showing the text and its subsequent tag

We can also display the tags from a visual perspective within Jupyter Notebook using the `render` function:

```
from spacy import displacy
displacy.render(nlp(str(sentences)), jupyter=True, style='ent')
```

Upon executing this code, we will receive the original paragraph, which has been color-coded based on the identified entity tags, allowing us to view the results visually:

[Biotechnology is a broad area of biology, involving the use of living systems and organisms to develop or make products., Depending on the tools and applications, it often overlaps with related scientific fields., In   the late 20th and early 21st centuries **DATE** . biotechnology has expanded to include new and diverse sciences, such as genomics, recombinant gene techniques, applied immunology, and development of pharmaceutical therapies and diagnostic tests., The term biotechnology was first used by   Karl Ereky **PERSON**   in   1919 **DATE** , meaning the production of products from raw materials with the aid of living organisms.]

Figure 9.9 – Results of the NER model when rendered in Jupyter Notebook

We can also use SciPy to implement a visual understanding of our text when it comes to parts of speech using the same `render()` function:

```
displacy.render(nlp(str(sentences[0])), style='dep', jupyter =
True, options = {'distance': 120})
```

We can see the output of this command in the following diagram:

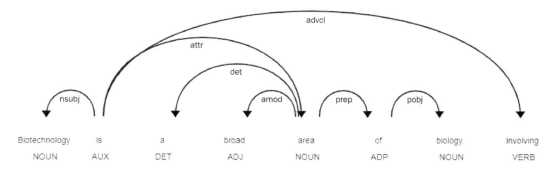

Figure 9.10 – Results of the POS model when rendered in Jupyter Notebook

This gives us a great way to understand and visualize the structure of a sentence before implementing any NLP models. With some of the basic analysis out of the way, let's go ahead and explore some applications of NLP using structured data.

Working with structured data

Now that we have explored some of the basics of NLP, let's dive into some more complex and common use cases that are often observed in the biotech and life sciences fields. When working with text-based data, it is much more common to work with larger datasets rather than single strings. More often than not, we generally want these datasets to involve scientific data regarding specific areas of interest relating to a particular research topic. Let's go ahead and learn how to retrieve scientific data using Python.

Searching for scientific articles

To programmatically retrieve scientific publication data using Python, we can make use of the `pymed` library from *PubMed* (`https://pubmed.ncbi.nlm.nih.gov/`). Let's go ahead and build a sample dataset:

1. First, let's import our libraries and instantiate a new `PubMed` object:

    ```
    from pymed import PubMed
    pubmed = PubMed()
    ```

2. Next, we will need to define and run our query. Let's go ahead and search for all the items related to monoclonal antibodies and retrieve `100` results:

    ```
    query = "monoclonal antibody"
    results = pubmed.query(query, max_results=100)
    ```

3. With the results found, we can iterate over our results to retrieve all the available fields for each of the given articles:

    ```
    articleList = []
    for article in results:
        articleDict = article.toDict()
        articleList.append(articleDict)
    ```

4. Finally, we can go ahead and convert our list into a DataFrame for ease of use:

    ```
    df = pd.DataFrame(articleList)
    df.head()
    ```

The following is the output:

	pubmed_id	title	abstract	keywords	journal	publication_date	authors	methods	conclusions	results	copyrights	doi	xml
0	34454387	Biodegradation of weathered crude oil by micro...	Oil spilled in the Arctic may drift into ice-c...	[Bacterial dynamics, Biotransformation, Brine...	Marine pollution bulletin	2021-08-29	[{'lastname': 'Lofthus' 'firstname': Synnøve...	None	None	None	Copyright © 2021 The Authors. Published by Els...	10.1016/j.marpolbul.2021.112823	[[]], [<Element Year at 0x000001FD2C569EA0>,...
1	34454204	Assessment of diagnostic potential of some cir...	Numerous studies have been carried out to iden...	[ALS, miR-142-3p, miR-143-3p, miR-206, miR-4516]	Clinical neurology and neurosurgery	2021-08-29	[{'lastname': 'Soliman' 'firstname': Radwa'...	None	This is the first study investigating miRNA pr...	As compared to the control group, significant...	Copyright © 2021 Published by Elsevier B.V.	10.1016/j.clineuro.2021.106883	[[]], [<Element Year at 0x000001FD2C57CD60>,...
2	34454155	Temperature influences the content and biosynt...	Temperature may affect the production of saxit...	[Alexandrium pacificum, HPLC-FLD, Saxitoxin bi...	The Science of the total environment	2021-08-29	[{'lastname': 'Wang' 'firstname': 'Hui' ini...	None	None	None	Copyright © 2021 Elsevier B.V. All rights rese...	10.1016/j.scitotenv.2021.149801	[[]], [<Element Year at 0x000001FD2C582810>,...
3	34454144	Towards understanding the link between the det...	The gradual degradation of technical materials...	[Adaptation mechanisms, Aerial green algae, Bi...	The Science of the total environment	2021-08-29	[{'lastname': 'Nowicka-Krawczyk', 'firstname'...	None	None	None	Copyright © 2021 Elsevier B.V. All rights rese...	10.1016/j.scitotenv.2021.149856	[[]], [<Element Year at 0x000001FD2C588E00>,...
4	34454141	A review on co-culturing of microalgae: A gree...	There is a growing global recognition that mic...	[Biomass, Co-cultivation, Lipid, Microalgae-ba...	The Science of the total environment	2021-08-29	[{'lastname': 'Ray', 'firstname': 'Ayushmita'...	None	None	None	Copyright © 2021 Published by Elsevier B.V.	10.1016/j.scitotenv.2021.149765	[[]], [<Element Year at 0x000001FD2C591360>,...

Figure 9.11 – A DataFrame showing the results of the PubMed search

With the final step complete, we have a dataset full of scientific abstracts and their associated metadata. In the next section, we will explore this data in more depth and develop a few visuals to represent it.

Exploring our datasets

Now that we have some data to work with, let's go ahead and explore it. If you recall from many of the previous chapters, we often explore our numerical datasets in various ways. We can group columns, explore trends, and find correlations – tasks we cannot necessarily do when working with text. Let's implement a few NLP methods to explore data in a slightly different way.

Checking string lengths

Since we have our dataset structured within a pandas DataFrame, one of the first items we generally want to explore is the distribution of string lengths for our text-based data. In the current dataset, the two main columns containing text are `title` and `abstract` – let's go ahead and plot the distribution of lengths:

```
sns.displot(df.abstract.str.len(), bins=25)
sns.displot(df.title.str.len(), bins=25)
```

The following is the output:

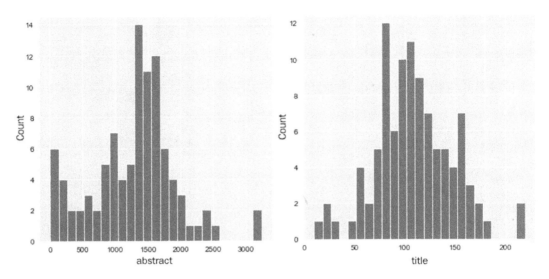

Figure 9.12 – Frequency distributions of the average length of abstracts (left) and titles (right)

Here, we can see that the average length of most abstracts is around 1,500 characters, whereas titles are around 100. Since the titles may contain important keywords or identifiers for a given article, similar to that of the abstracts, it would be wise to combine the two into a single column to analyze them together. We can simply combine them using the + operator:

```
df["text"] = df["title"] + " " + df["abstract"]
df[["title", "abstract", "text"]]
```

We can see this new column in the following screenshot:

	title	abstract	text
0	Protective effects of anti-HMGB1 monoclonal an...	During ischemia reperfusion (IR) injury, high ...	Protective effects of anti-HMGB1 monoclonal an...
1	Current pharmacological approaches and potenti...	Celiac Disease (CeD) is estimated to currently...	Current pharmacological approaches and potenti...
2	Detailed analysis of anti-emicizumab antibody ...	Emicizumab is a humanized bispecific monoclona...	Detailed analysis of anti-emicizumab antibody ...
3	Neuroprotective Effects of Anti-high Mobility ...	High mobility group box-1 (HMGB1) is a ubiquit...	Neuroprotective Effects of Anti-high Mobility ...
4	Potential and pitfalls of	The relation between tumor uptake and target c...	Potential and pitfalls of The relation betwee...

Figure 9.13 – A sample DataFrame showing the title, abstract, and text columns

Using the mean() function on each of the columns, we can see that the titles have, on average, 108 characters, the abstracts have 1,277 characters, and the combined text column has 1,388 characters.

Similar to other datasets, we can use the `value_counts()` function to get a quick sense of the most common words:

```
df.text.str.split(expand=True).stack().value_counts()
```

Immediately, we notice that our dataset is flooded with stop words:

```
of      786
the     712
and     643
in      447
to      328
```

Figure 9.14 – A sample of the most frequent words in the dataset

We can implement the same `cleaner` function as we did previously to remove these stop words and any other undesirable values. Note that some of the cells within the DataFrame may be empty, depending on the query made and the results returned. We'll take a closer look at this in the next section.

Cleaning text data

We can add a quick check at the top of the function by checking the type of the value to ensure that no errors are encountered:

```
from nltk.corpus import stopwords
STOP_WORDS = stopwords.words()


def cleaner(text):
    if type(text) == str:
        text = text.lower()
        text = re.sub("[^a-zA-Z]+", ' ', text)
        text_tokens = word_tokenize(text)
        tokens_without_sw = [word for word in text_tokens if
not word in STOP_WORDS]
        filtered_sentence = (" ").join(tokens_without_sw)
        return filtered_sentence
```

We can quickly test this out on a sample string to test the functionality:

```
cleaner("Biotech in 2021 is a wonderful field to work and study
in!")
```

The output of this function can be seen in the following screenshot:

```
'biotech wonderful field work study'
```

Figure 9.15 – Results of the cleaning function

With the function working, we can go ahead and apply this to the `text` column within the DataFrame and create a new column consisting of the cleaned text using the `apply()` function, which allows us to apply a given function iteratively down through all rows of a DataFrame:

```
df["clean_text"] = df["text"].apply(lambda x: cleaner(x))
```

We can check the performance of our function by checking the columns of interest:

```
df[["text", "clean_text"]].head()
```

We can see these two columns in the following screenshot:

	text	clean_text
0	Protective effects of anti-HMGB1 monoclonal an...	protective effects anti hmgb monoclonal antibo...
1	Current pharmacological approaches and potenti...	current pharmacological approaches potential f...
2	Detailed analysis of anti-emicizumab antibody ...	detailed analysis anti emicizumab antibody dec...
3	Neuroprotective Effects of Anti-high Mobility ...	neuroprotective effects anti high mobility gro...
4	Potential and pitfalls of The relation betwee...	potential pitfalls relation tumor uptake targe...

Figure 9.16 – A DataFrame showing the original and cleaned texts

If we go ahead and check `value_counts()` of the `clean_text` column, as we did previously, you will notice that the stop words were removed and that more useful keywords are now populated at the top.

Creating word clouds

Another popular and useful method to get a quick sense of the content for a given text-based dataset is by using word clouds. **Word clouds** are images that are populated with the content of a dataset in which words are rendered in larger fonts when they are more frequent, and smaller fonts when they are less frequent. To accomplish this, we can make use of the wordclouds library:

1. First, we need to import the function and then create a wordcloud object that we will specify a number of parameters in. We can adjust the dimensions of the image, colors, and the data as follows:

```
from wordcloud import WordCloud, STOPWORDS
plt.figure(figsize=(20,10))
# Drop nans
df2 = df[["clean_text"]].dropna()
# Create word cloud
wordcloud = WordCloud(width = 5000,
                      height = 3000,
                      random_state=1,
                      background_color='white',
                      colormap='Blues',
                      collocations=False,
                      stopwords = STOPWORDS).generate('
'.join(df2['clean_text']))
```

2. Now, we can use the imshow() function from matplotlib to render the image:

```
plt.figure( figsize=(15,10) )
plt.imshow(wordcloud)
```

The following is the output:

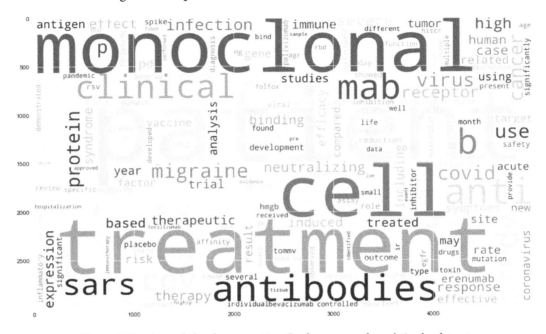

Figure 9.17 – A word cloud representing the frequency of words in the dataset

In this section, we investigated a few of the most popular methods for quickly analyzing text-based data as a preliminary step before performing any type of rigorous analysis or model development process. In the next section, we will train a model using this dataset to investigate topics.

Tutorial – clustering and topic modeling

Similar to some of the previous examples we have seen so far, much of our data can either be classified in a supervised setting or clustered in an unsupervised one. In most cases, text-based data is generally made available to us in the form of real-world data in the sense that it is in a raw and unlabeled form.

Let's look at an example where we can make sense of our data and label it from an unsupervised perspective. Our main objective here will be to preprocess our raw text, cluster the data into five clusters, and then determine the main topics for each of those clusters. If you are following along using the provided code and documentation, please note that your results may vary as the dataset is dynamic, and its contents change as new data is populated into the PubMed database. I would urge you to customize the queries to topics that interest you. With that in mind, let's go ahead and begin.

We will begin by querying some data using the pymed library and retrieving a few hundred abstracts and titles to work with:

```
def dataset_generator(query, num_results, ):
    results = pubmed.query(query, max_results=num_results)
    articleList = []
    for article in results:
        articleDict = article.toDict()
        articleList.append(articleDict)
    print(f"Found {len(articleList)} results for the query
'{query}'.")
    return pd.DataFrame(articleList)
```

Instead of making a single query, let's make a few and combine the results into a single DataFrame:

```
df1 = dataset_generator("monoclonal antibodies", 600)
df2 = dataset_generator("machine learning", 600)
df3 = dataset_generator("covid-19", 600)
df4 = dataset_generator("particle physics", 600)


df = pd.concat([df1, df2, df3, df4])
```

Taking a look at the data, we can see that some of the cells have missing (nan) values. Given that our objective concerns the text-based fields only (titles and abstracts), let's limit the scope of any cleaning methods to those columns alone:

```
df = df[["title", "abstract"]]
```

Given that we are concerned with the contents of each article as a whole, we can combine the titles and abstracts together into a new column called text:

```
df["text"] = df["title"] + " " + df["abstract"]
df = df.dropna()
print(df.shape)
```

Taking a look at the dataset, we can see that we have 560 rows and 3 columns. Scientific articles can be very descriptive, encompassing many stop words. Given that our objective here is to detect topics that are represented by keywords, let's remove any punctuation, numerical values, and stopwords from our text:

```
def cleaner(text):
    if type(text) == str:
        text = text.lower()
        text = re.sub("[^a-zA-Z]+", ' ', text)
        text_tokens = word_tokenize(text)
        tokens_without_sw = [word for word in text_tokens if
not word in STOP_WORDS]
        filtered_sentence = (" ").join(tokens_without_sw)
        return filtered_sentence
df["text"] = df["text"].apply(lambda x: cleaner(x))
```

We can check the average number of words per article before and after implementing the script to ensure that the data was cleaned. In our case, we can see that we started with an average of 190 words and ended up with an average of 123.

With the data now clean, we can go ahead and extract our features. We will use a relatively simple and common method known as **TFIDF** – a measure of originality of words in which each word is compared to the number of times it appears in an article, relative to the number of articles the same word appears in. We can think of TFIDF as two separate items – **Term Frequency (TF)** and **Inverse Document Frequency (IDF)** – which we can represent as follows:

$$TF * IDF = TF(t,d) * IDF(t)$$

In the preceding equation, t is the term or keyword, while d is the document or – in our case – the article. The main idea here is to capture important keywords that would be descriptive as main topics but ignore those that appear in almost every article. We will begin by importing `TfidfVectorizer` from `sklearn`:

```
from sklearn.feature_extraction.text import TfidfVectorizer
```

Next, we will convert our text-based data into numerical features by fitting our dataset and transforming the values:

```
vectors = TfidfVectorizer(stop_words="english", max_
features=5500)
```

```
vectors.fit(df.text.values)
features = vectors.transform(df.text.values)
```

We can check the shape of the `features` variable to confirm that we have 560 rows, just as we did before applying TFIDF, and 5,500 columns worth of features to go with it. Next, we can go ahead and cluster our documents using one of the many clustering methods we have explored so far.

Let's implement `MiniBatchKMeans` and specify 4 as the number of clusters we want to retrieve:

```
from sklearn.cluster import MiniBatchKMeans
cls = MiniBatchKMeans(n_clusters=4)
cls.fit(features)
```

When working with larger datasets, especially in production, it is generally advisable to avoid using pandas DataFrames as there are more efficient methods available, depending on the processes you need to implement. Given that we are only working with 560 rows of data, and our objective is to cluster our data and retrieve topics, we will once again make use of DataFrames to manage our data. Let's go ahead and add our predicted clusters to our DataFrame:

```
df["cluster"] = cls.predict(features)
df[["text", "cluster"]].head()
```

We can see the output of this command in the following screenshot:

	text	cluster
0	development chemiluminescence immunoassay accu...	0
1	mesenchymal stem cells sars cov infection hype...	1
2	peroxidase mimicking nanozyme surface disperse...	0
3	simultaneous engagement tumor stroma targeting...	4
4	ordinary proteins adsorption molecular orienta...	0

Figure 9.18 – A DataFrame showing the cleaned texts and their associated clusters

With the data clustered, let's plot this data in a 2D scatterplot. Given that we have several thousand features, we can make use of the **PCA** algorithm to reduce these down to only two features for our visualization:

```
from sklearn.decomposition import PCA
pca = PCA(n_components=2)
pca_features_2d = pca.fit_transform(features.toarray())
pca_features_2d_centers = pca.transform(cls.cluster_centers_)
```

Let's go ahead and add these two principal components to our DataFrame:

```
df["pc1"], df["pc2"] = pca_features_2d[:,0], pca_
features_2d[:,1]
df[["text", "cluster", "pc1", "pc2"]].head()
```

We can see the output of this command in the following screenshot:

	text	cluster	pc1	pc2
0	development chemiluminescence immunoassay accu...	3	0.036873	-0.031256
1	mesenchymal stem cells sars cov infection hype...	1	0.145596	0.116410
2	peroxidase mimicking nanozyme surface disperse...	2	-0.073850	-0.091979
3	simultaneous engagement tumor stroma targeting...	3	0.009339	-0.085552
4	ordinary proteins adsorption molecular orienta...	2	-0.056036	-0.120671

Figure 9.19 – A DataFrame showing the texts, clusters, and principal components

Here, we can see that each row of text now has a cluster, as well as a set of coordinates, in the form of principal components. Next, we will plot our data and color by cluster:

```
plt.figure(figsize=(15,8))
new_cmap = matplotlib.colors.LinearSegmentedColormap.from_
list("mycmap", colors)
plt.scatter(df["pc1"], df["pc2"], c=df["cluster"], cmap=new_
cmap)
plt.scatter(pca_features_2d_centers[:, 0], pca_features_2d_
centers[:,1], marker='*', s=500, c='r')
plt.xlabel("PC1", fontsize=20)
plt.ylabel("PC2", fontsize=20)
```

Upon executing this code, we will get the following output:

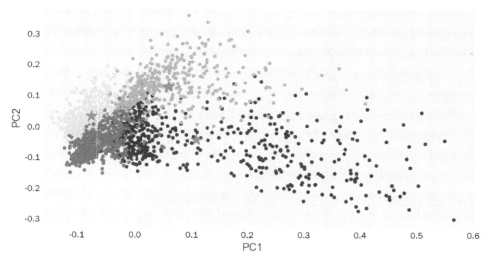

Figure 9.20 – A scatterplot of the principal components colored by cluster, with stars representing the cluster centers

Here, we can see that there seems to be some adequate separation between clusters! The two clusters on the far left seem to have smaller variance in their distributions, whereas the other two are much more spread out. Given that we reduced a considerable number of features down to only two principal components, it makes sense that there is a certain degree of overlap between them, especially given the fact that all the articles were scientific.

Now, let's go ahead and calculate some of the most prominent topics that were found in our dataset:

1. First, we will begin by implementing TFIDF:

```
from sklearn.feature_extraction.text import
CountVectorizer, TfidfVectorizer
vectors = TfidfVectorizer(max_features=5500, stop_
words="english")
nmf_features = vectors.fit_transform(df.text)
```

2. Next, we will reduce the dimensionality. However, this time, we will use **Non-Negative Matrix Factorization (NMF)** to reduce our data instead of PCA. We will need to specify the number of topics we are interested in:

```
from sklearn.decomposition import NMF
n_topics = 10
```

```
cls = NMF(n_components=n_topics)
cls.fit(features)
```

3. Now, we can specify the number of keywords to retrieve per topic. After that, we will iterate over the components and retrieve the keywords of interest:

```
num_topic_words = 3
feature_names = vectors.get_feature_names()
for i, j in enumerate(cls.components_):
    print(i, end=' ')
    for k in j.argsort()[-1:-num_topic_words-1:-1]:
        print(feature_names[k], end=' ')
```

Upon executing this loop, we retrieve the following as output:

```
0 cov sars variants
1 learning model machine
2 covid disease coronavirus
3 patients treatment treated
4 particle particles energy
5 antibodies antibody monoclonal
6 vaccine vaccination vaccines
7 cancer cells cell
8 data review artificial
9 pandemic covid health
```

Figure 9.21 – Top 10 topics of the dataset, each represented by three keywords

We can use these topic modeling methods to extract insights and trends from the dataset, allowing users to have high-level interpretations without the need to dive into the datasets as a whole. Throughout this tutorial, we examined one of the classical methods for clustering and topic modeling: using **TFIDF** and **NMF**. However, many other methods exist that use language models, such as **BERT** and **BioBERT** and libraries such as **Gensim** and **LDA**. If this is an area you find interesting, I highly urge you to explore these libraries for more information.

Often, you will not have your data already existing in a usable format. In this tutorial, we had our dataset structured within a DataFrame, ready for use to slice and dice. However, in many cases, our data of interest will be unstructured, such as in PDFs. We will explore how to handle situations such as these in the next section.

Working with unstructured data

In the previous section, we explored some of the most common tasks and processes that are conducted when handing text-based data. More often than not, you will find that the data you work with is generally not of a structured nature, or perhaps not of a digital nature. Take, for example, a company that has decided to move all printed documents to a digital state. Or perhaps a company that maintains a large repository of documents, none of which are structured or organized. For tasks such as these, we can rely on several AWS products to come to our rescue. We will explore two of the most useful NLP tools in the next few sections.

OCR using AWS Textract

In my opinion, one of the most useful tools available within **AWS** is an **Optical Character Recognition (OCR)** tool known as **AWS Textract**. The main idea behind this tool is to enable users to extract text, tables, and other useful items from images or static PDF documents using pre-built machine learning models implemented within Textract.

For example, users can upload images or scanned PDF documents to Textract that are otherwise unsearchable and extract all the text-based content from them, as shown in the following diagram:

Figure 9.22 – A schematic showing structuring raw PDFs into organized digital text

In addition to extracting text, users can extract key-value pairs such as those found in both printed and handwritten forms:

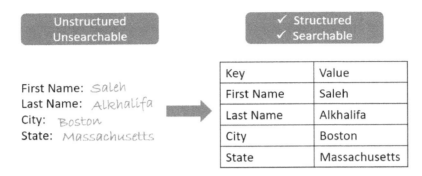

Figure 9.23 – A schematic showing the structuring of handwritten data to organized tables

To use **Textract** in Python, we will need to use an AWS Python library known as `boto3`. We can install Boto3 using `pip`. When using Boto3 within SageMaker, you will not be required to use any access keys to utilize the service. However, if you are using a local implementation of Jupyter Notebook, you will need to be able to authenticate using access keys. **Access keys** can easily be created in a few simple steps:

1. Navigate to the AWS console and select **IAM** from the **Services** menu.

2. Under the **Access Management** tab on the left, click on **Users**, then **Add Users**.

3. Go ahead and give your user a name, such as `ml-biotech-user`, and enable the **Programmatic access** option:

Set user details

You can add multiple users at once with the same access type and permissions. Learn more

User name*   ml-biotech-user

○ Add another user

Select AWS access type

Select how these users will access AWS. Access keys and autogenerated passwords are provided in the last step. Learn more

Access type*   ☑ **Programmatic access**
Enables an **access key ID** and **secret access key** for the AWS API, CLI, SDK, and other development tools.

☐ **AWS Management Console access**
Enables a **password** that allows users to sign-in to the AWS Management Console.

Figure 9.24 – Setting the username for AWS IAM roles

4. Next, select the **Attach existing policies directly** option at the top and add the policies of interest. Go ahead and add Textract, Comprehend, and S3 as we will require all three for the role:

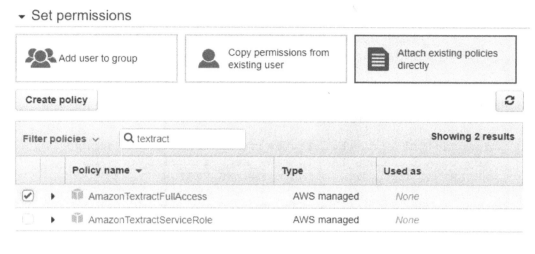

Figure 9.25 – Setting the policies for AWS IAM roles

5. After labeling your new user with some descriptive tags, you will gain access to two items: your **access key ID** and **AWS secret access k**. Be sure to copy those two items to a safe space. For security, you will not be able to retrieve them from AWS after leaving this page.

Now that we have our access keys, let's go ahead and start implementing Textract on a document of interest. We can complete this in a few steps.

1. First, we will need to upload our data to our **S3 bucket**. We can recycle the same S3 bucket we used earlier in this book. We will need to specify our keys, and then connect to AWS using the Boto3 client:

```
AWS_ACCESS_KEY_ID = "add-access-key-here"
AWS_SECRET_ACCESS_KEY = "add-secret-access-key-here"
AWS_REGION = "us-east-2"


s3_client = boto3.client('s3', aws_access_key_id=AWS_
ACCESS_KEY_ID, aws_secret_access_key=AWS_SECRET_ACCESS_
KEY, region_name=AWS_REGION)
```

2. With our connection set up, we can go ahead and upload a sample PDF file. Note that you can submit PDFs as well as image files (PNG). Let's go ahead and upload our PDF file using the `upload_fileobj()` function:

```
with open("Monoclonal Production Article.pdf", "rb") as
f:
    s3_client.upload_fileobj(f, "biotech-machine-
learning", "pdfs/Monoclonal Production Article.pdf")
```

3. With our PDF now uploaded, we can use Textract. First, we will need to connect using the Boto3 client. Note that we changed the desired resource from `'s3'` to `'textract'` since we are using a different service now:

```
textract_client = boto3.client('textract', aws_access_
key_id=AWS_ACCESS_KEY_ID, aws_secret_access_key=AWS_
SECRET_ACCESS_KEY, region_name=AWS_REGION)
```

4. Next, we can send our file to Textract using the `start_document_text_detection()` method, where we specify the name of the bucket and the name of the document:

```
response = textract_client.start_document_text_detection(
                DocumentLocation={'S3Object':
{'Bucket': "biotech-machine-learning", 'Name': "pdfs/
Monoclonal Production Article.pdf"} })
```

5. We can confirm that the task was started successfully by checking the status code in the response variable. After a few moments (depending on the duration of the job), we retrieve the results by specifying `JobId`:

```
results = textract_client.get_document_text_
detection(JobId=response["JobId"])
```

Immediately, we will notice that the `results` variable is simply one large JSON that we can parse and iterate over. Notice that the structure of the JSON is quite complex and detailed.

6. Finally, we can gather all the text by iterating over `Blocks` and collecting all the texts for blocks of the `LINE` type:

```
documentText = ""
for item in results["Blocks"]:
    if item["BlockType"] == "LINE":
        documentText = documentText + item["Text"]
```

If you print the documentText variable, you will see all of the text that was successfully collected from that document! Textract can be an extremely useful tool for moving documents from an unstructured and unsearchable state to a more structured and searchable state. Often, most text-based data will exist in an unstructured format, and you will find Textract to be one of the most useful resources for these types of applications. Textract is generally coupled with other AWS resources to maximize the utility of the tool, such as **DynamoDB** for storage or **Comprehend** for analysis. We will explore Comprehend in the next section.

Entity recognition using AWS Comprehend

Earlier in this chapter, we implemented a NER model using the SciPy library to detect entities in a given section of text. Now, let's explore a more powerful implementation of NER known as **AWS Comprehend**. Comprehend is an NLP service that's designed to discover insights in unstructured text data, allowing users to extract key phrases, calculate sentiment, identify entities, and much more. Let's go ahead and explore this tool.

Similar to other **AWS** resources, we will need to connect using the boto3 client:

```
comprehend_client = boto3.client('comprehend', aws_access_key_
id=AWS_ACCESS_KEY_ID, aws_secret_access_key=AWS_SECRET_ACCESS_
KEY, region_name=AWS_REGION)
```

Next, we can go ahead and use the detect_entities() function to identify entities in our text. We can use the documentText string we generated using Textract earlier:

```
response = comprehend_client.detect_entities(
    Text=documentText[:5000],
    LanguageCode='en',
)
print(response["Entities"])
```

Upon printing the response, we will see the results for each of the entities that were detected in our block of text. In addition, we can organize the results in a DataFrame:

```
pd.DataFrame(response["Entities"]).sort_values(by='Score',
ascending=False).head()
```

Upon sorting the values by score, we can see our results listed in a structured manner:

	Score	Type	Text	BeginOffset	EndOffset
8	0.997775	DATE	1 May 1973	248	258
17	0.991282	ORGANIZATION	National Institute of Allergy and Infectious	2798	2842
15	0.981901	ORGANIZATION	U.S.Public Health Service	2678	2703
19	0.966133	PERSON	N. R. Klinman	2870	2883
2	0.957784	PERSON	NORMAN R. KLINMAN	107	124

Figure 9.26 – A sample DataFrame showing the results of the AWS Comprehend entities API

In addition to entities, Comprehend can also detect key phrases within text:

```
response = comprehend_client.detect_key_phrases(
    Text=documentText[:5000],
    LanguageCode='en',
)
response["KeyPhrases"][0]
```

Upon printing the first item in the list, we can see the score, phrase, and position in the string:

```
{'Score': 0.94112229347229,
 'Text': 'Brief Definitive ReportsMONOCLONAL PRODUCTION',
 'BeginOffset': 0,
 'EndOffset': 45}
```

Figure 9.27 – Results of the AWS Comprehend key phrases API

In addition, we can also detect the sentiment using the detect_sentiment() function:

```
response = comprehend_client.detect_sentiment(
    Text=documentText[:5000],
    LanguageCode='en',
)
print(response)
```

We can print the response variable to get the results of the string. We can see that the sentiment was noted as neutral, which makes sense for a statement concerning scientific data that is generally not written with a positive or negative tone:

```
{'Sentiment': 'NEUTRAL',
 'SentimentScore': {'Positive': 0.0010422870982438326,
  'Negative': 0.00028130708960816264,
  'Neutral': 0.9986714124679565,
  'Mixed': 4.950157290295465e-06},
 'ResponseMetadata': {'RequestId': 'ede1155b-dbf4-41f5-a9c8-33d498b61c04',
  'HTTPStatusCode': 200,
  'HTTPHeaders': {'x-amzn-requestid': 'ede1155b-dbf4-41f5-a9c8-33d498b61c04',
   'content-type': 'application/x-amz-json-1.1',
   'content-length': '166',
   'date': 'Tue, 24 Aug 2021 01:54:35 GMT'},
  'RetryAttempts': 0}}
```

Figure 9.28 – Results of the AWS Comprehend sentiment API

Lastly, Comprehend can also detect dominant languages within text using the `detect_dominant_language()` function:

```
response = comprehend_client.detect_dominant_language(
    Text=documentText[:5000],
)
response
```

Here, we can see that, upon printing the response, we get a sense of the language, as well as the associated score or probability from the model:

```
{'Languages': [{'LanguageCode': 'en', 'Score': 0.9832875728607178}],
 'ResponseMetadata': {'RequestId': 'f8f874a8-7631-4e4e-88d6-345348f716fd',
  'HTTPStatusCode': 200,
  'HTTPHeaders': {'x-amzn-requestid': 'f8f874a8-7631-4e4e-88d6-345348f716fd',
   'content-type': 'application/x-amz-json-1.1',
   'content-length': '64',
   'date': 'Tue, 24 Aug 2021 01:55:23 GMT'},
  'RetryAttempts': 0}}
```

Figure 9.29 – Results of the AWS Comprehend language detection API

AWS Textract and AWS Comprehend are two of the top NLP tools available today and have been instrumental in structuring and analyzing vast amounts of unstructured text documents. Most NLP-based applications today generally use at least one, if not both, of these types of technologies. For more information about Textract and Comprehend, I highly recommend that you visit the AWS website (`https://aws.amazon.com/`).

So far, we have learned how to analyze and transform text-based data, especially when it comes to moving data from an unstructured state to a more structured state. Now that the documents are more organized, the next step is to be able to use them in one way or another, such as through a search engine. We will learn how to create a **semantic search** engine using **transformers** in the next section.

Tutorial – developing a scientific data search engine using transformers

So far, we have looked at text from a word-by-word perspective in the sense that we kept our text *as is*, without the need to convert or embed it in any way. In some cases, converting words into numerical values or **embeddings** can open many new doors and unlock many new possibilities, especially when it comes to deep learning. Our main objective within this tutorial will be to develop a search engine to find and retrieve scientific data. We will do so by implementing an important and useful deep learning NLP architecture known as a transformer. The main benefit here is that we will be designing a powerful semantic search engine in the sense that we can now search for ideas or semantic meaning rather than only keywords.

We can think of transformers as deep learning models designed to solve sequence-based tasks using a mechanism known as **self-attention**. We can think of self-attention as a method to help relate different portions of text within a sentence or embedding in an attempt to create a representation. Simply put, the model attempts to view sentences as ideas, rather than a collection of single words.

Before we begin to work with transformers, let's talk a little more about the idea of **embeddings**. We can think of embeddings as low-dimensional numerical values or vectors of continuous numbers representing an item, which in our case would be a word or sentence. We commonly convert words and sentences into embeddings to allow models to carry out machine learning tasks more easily when working with larger datasets. Within the context of NLP and neural networks, there are three main reasons embeddings are used:

- To reduce the **dimensionality** of large segments of text data
- To compute the **similarity** between two different texts
- To **visualize** relationships between portions of text

Now that we have gained a better sense of embeddings and the role they play in NLP, let's go ahead and get started with a real-world example of a scientific search engine. We will begin by importing a few libraries that we will need:

```
import scipy
import torch
import pandas as pd
from sentence_transformers import SentenceTransformer, util
```

To create our embeddings, we will need a model. We have the option to create a customized model concerning our dataset. The benefit here is that our results would likely improve, given that the model was trained on text about our domain. Alternatively, we could use other pre-trained models available from the SentenceTransformer website (https://www.sbert.net/). Let's download one of these pre-trained models:

```
model = SentenceTransformer('msmarco-distilbert-base-v4')
```

Next, we can create a testing database and populate it with a few sentences:

```
database = df["abstract"].values
```

Next, we can call the encode() function to convert our list of strings into a list of embeddings:

```
database_embeddings = model.encode(database)
```

If we check the length of the database and the length of database_embeddings using the len() function, we will find that they both contain the same number of elements since there should be one embedding for every piece of text. If we print the contents of the first element of the embeddings database, we will find that the content is now simply a list of vectors:

```
array([ 8.95956755e-02,  2.89472520e-01,  3.39112878e-02, -9.27880332e-02,
        1.26130179e-01,  1.55073792e-01, -1.76565468e-01, -3.21197420e-01,
        1.10137619e-01, -3.98256928e-01,  1.44913137e-01, -9.81106758e-02,
       -6.49801344e-02,  1.24519557e-01,  1.55980423e-01, -1.48738772e-01,
```

Figure 9.30 – A view of an embedded piece of text

With each of our documents now embedded, the idea would be that a user would want to search for or query a particular phrase. We can take a user's query and encode it as we did with the others, but assign that value to a new variable that we will call query_embedding:

```
query = "One of the best discoveries were monoclonal
antibodies"
query_embedding = model.encode(query)
```

With the query and sentences embedded, we can compute the distance between the items. The idea here is that documents that were more similar to the user's query would have shorter distances, and those that were less similar would have longer ones. Notice that we are using cosine here as a measure of distance, and therefore, similarity. We can use other methods as well, such as the euclidean distance:

```
import scipy
cos_scores = util.pytorch_cos_sim(query_embedding,
                        database_embeddings)[0]
```

Let's go ahead and prepare a single runSearch function that incorporates the query, the encoder, as well as a method to display our results. The process begins with a few print statements, and then encodes the new query into a variable called query_embedding. The distances are then calculated, and the results are sorted according to their distance. Finally, the results are iterated over and the scores, titles, and abstracts for each are printed:

```
def askQuestion(query, top_k):
    print(f"###################################")
    print(f"#### {query} ####")
    print(f"###################################")
    query_embedding = model.encode(query, convert_to_
tensor=True)
    cos_scores = util.pytorch_cos_sim(query_embedding,
                database_embeddings)[0]
    top_results = torch.topk(cos_scores, k=top_k)

    for score, idx in zip(top_results[0], top_results[1]):
        print("#### Score: {:.4f}".format(score))
        print("#### Title: ", df.loc[float(idx)].title)
        print("#### Abstract: ", df.loc[float(idx)].abstract)
```

```
print("###################################")
```

Now that we have prepared our function, we can call it with our query of interest:

```
query = ' What is known about the removal of harmful
cyanobacteria?
askQuestion(query, 5)
```

Upon calling the function, we retrieve several results printed similarly. We can see one of the results in the following screenshot, showing us the `score`, `title`, and `abstract` properties of the article:

```
########################################################################################
#### Removal of harmful cyanobacteria ####
########################################################################################
#### Score: 0.6420
#### Title: Simultaneous removal of colonial Microcystis and microcystins by protozoa grazing coupled with ultrasound treatment.
#### Abstract: Removal of harmful cyanobacteria is an extremely urgent task in global lake management and protection. Convention
al measures are insufficient for simultaneously removing cyanobacteria and hazardous cyanotoxin, efficient and environmental-frie
ndly measures are therefore particularly needed. Herbivorous protozoa have great potentials in controlling algae, however, large-
sized colonial Microcystis is inedible for protozoa, which is a central problem to be solved. Therefore, in present study, a meas
ure of protozoa grazing assisted by ultrasound was investigated in laboratory scale for eliminating harmful colonial Microcystis.
The results showed that with ultrasound power and time increasing, the proportion of unicellular Microcystis increased significan
tly. With Ochromonas addition, approximately 80% of colonial Microcystis and microcystin was removed on day 4 under ultrasound po
wer of 100 W for 15 min, while Ochromonas only reduced Microcystis by less than 20% without assistance of ultrasound. Moreover, w
hen directly exposed to low-intensity ultrasound, Ochromonas showed strong resistance to ultrasound and were not inhibited in gra
zing Microcystis. Overall, ultrasound increases edible food for protozoa via collapsing Microcystis colonies and assists Ochromon
as to remove Microcystis, thus intermittently collapsing colonial Microcystis using low-intensity ultrasound can significantly im
prove the removal efficiency of Microcystis by protozoa grazing, which provided a new insight in controlling harmful colonial Mic
rocystis.
```

Figure 9.31 – Results of the semantic searching model for scientific text

With that, we have managed to successfully develop a semantic searching model capable of searching through scientific literature. Notice that the query itself is not a direct string match to the top result the model returned. Again, the idea here is not to match keywords but to calculate the distance between embeddings, which is representative of similarities.

Summary

In this chapter, we made an adventurous attempt to cover a wide range of NLP topics. We explored a range of introductory topics such as NER, tokenization, and parts of speech using the NLTK and spaCy libraries. We then explored NLP through the lens of structured datasets, in which we utilized the `pymed` library as a source for scientific literature and proceeded to analyze and clean the data in our preprocessing steps. Next, we developed a word cloud to visualize the frequency of words in a given dataset. Finally, we developed a clustering model to group our abstracts and a topic modeling model to identify prominent topics.

We then explored NLP through the lens of unstructured data in which we explored two common AWS NLP products. We used Textract to convert PDFs and images into searchable and structured text and Comprehend to analyze and provide insights. Finally, we learned how to develop a semantic search engine using deep learning transformers to find pertinent information.

What was particularly unique about this chapter is that we learned that text is a sequence-based type of data, which makes its uses and applications drastically different from many of the other datasets we had previously worked with. As companies across the world begin to migrate legacy documents into the digital space, the ability to search for documents and identify insights will be of great value. In the next chapter, we will examine another type of sequence-based data known as time series.

10
Exploring Time Series Analysis

In the previous chapter, we discussed using deep learning and its robust applicability when it comes to unstructured data in the form of natural language – a type of sequential data. Another type of sequential data that we will now turn our attention to is time series data. We can think of time series data as being standard datasets yet containing a time-based feature, thus unlocking a new set of possibilities when it comes to developing predictive models.

One of the most common applications in time series data is a process known as time series analysis. We can define time series analysis as an area of data **exploration** and **forecasting** in which datasets are ordered or indexed using a particular **time interval** or **timestamp**. There are many examples of time series data that we encounter in the biotechnology and life sciences industries daily. Some of the more laboratory-based areas of focus include gene expression and chromatography, as well as non-lab areas such as demand forecasting and stock price analysis.

Throughout this chapter, we will explore several different areas when it comes to gaining a better understanding of the analysis of time series data, as well as developing a model capable of consuming this data and developing a robust, predictive model.

As we explore these areas, we will cover the following topics:

- Understanding time series data
- Exploring the components of a time series dataset
- Tutorial – forecasting product demand using Prophet and LSTM

With that in mind, let's go ahead and get started!

Understanding time series data

When it comes to using **time series** data, there are endless ways to visualize and display data to effectively communicate a thought or idea. In most of the data we have used so far, we have handled features and labels in which a certain set of features generally corresponded to a label of interest. When it comes to time series data, we tend to forego the idea of a class or label and focus more on trends within the data instead. One of the most common applications of time series data is the idea of **demand forecasting**. Demand forecasting, as its name suggests, comprises the many methods and tools available to help predict demand for a given good or service ahead of time. Throughout this section, we will learn about the many aspects of time series analysis using a dataset concerning the demand forecasting of a given biotechnology product.

Treating time series data as a structured dataset

There are many different biotechnology products on the market today, ranging from agricultural genetically modified crops, all the way to monoclonal antibody therapeutics. In this section, we will investigate the sales data of a human therapeutic by using the `dataset_demand-forecasting_ts.csv` dataset, which belongs to a small biotech start-up:

1. With this in mind, let's go ahead and dive into the data. We will begin by importing the libraries of interest, importing the CSV file, and taking a glance at the first few rows of data:

```
import pandas as pd
df = pd.read_csv("dataset_demand-forecasting_ts.csv")
df.head()
```

This will result in the following output:

	Date	Sales
0	2014-01-01	11219
1	2014-01-02	12745
2	2014-01-03	10498
3	2014-01-04	12028
4	2014-01-05	13900

Figure 10.1 – The first few rows of the forecasting dataset

Relative to the many other datasets we have worked with in the past, this one seems much simpler in the sense that we are working with only two columns: Date and the number of Sales for any given day. We can also see that the sales have been aggregated by day, starting on 2014-01-01. If we check the end of the dataset using the tail() function, we will see that the dataset ends on 2020-12-23 – essentially providing us with 6 years' worth of sales data to work with.

2. We can visualize the time series data using the Plotly library:

```
import plotly.express as px
import plotly.graph_objects as go
fig = px.line(df, x="Date", y="Sales", title='Single
Product Demand', width=800, height=400)
fig.update_traces(line_color='#4169E1')
fig.show()
```

Upon executing the `fig.show()` function, we will receive the following output:

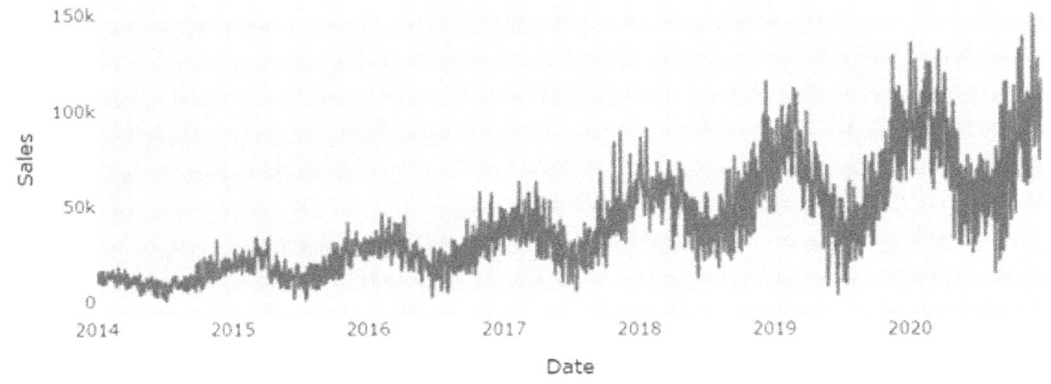

Figure 10.2 – Time series plot of the sales dataset

We can immediately make a few initial observations regarding the dataset:

- There is a significant amount of noise and variability within the data.

- The sales gradually increase over time (I should have invested in them!).

- There seems to be an element of seasonality in which sales peak around December.

To explore these ideas a bit more and dive deeper into the data, we will need to deconstruct the time series aspect. Using the `Date` column, we can break the dataset down into years, months, and days to get a better sense of the repetitive or **seasonal** nature of this data.

> **Important note**
>
> **Seasonality** within datasets refers to the seasonal characteristics relating to that time of the year. For example, datasets relating to the flu shot often show increased rates in the fall relative to the spring or summer in preparation for the winter (flu season).

3. First, we will need to use the `to_datetime()` function to convert `string` into the `date` type:

```
def get_features(dataframe):
    dataframe["sales"] = dataframe["sales"]
    dataframe["Date"] = pd.to_datetime(dataframe['Date'])
    dataframe['year'] = dataframe.Date.dt.year
    dataframe['month'] = dataframe.Date.dt.month
```

```
    dataframe['day'] = dataframe.Date.dt.day
    dataframe['dayofyear'] = dataframe.Date.dt.dayofyear
    dataframe['dayofweek'] = dataframe.Date.dt.dayofweek
    dataframe['weekofyear'] = dataframe.Date.
dt.weekofyear
    return dataframe


df = get_features(df)
df.head()
```

Upon executing this command, we will receive the following DataFrame as output:

	Date	sales	year	month	day	dayofyear	dayofweek	weekofyear
0	2014-01-01	11219	2014	1	1	1	2	1
1	2014-01-02	12745	2014	1	2	2	3	1
2	2014-01-03	10498	2014	1	3	3	4	1
3	2014-01-04	12028	2014	1	4	4	5	1
4	2014-01-05	13900	2014	1	5	5	6	1

Figure 10.3 – The first five rows of the sales dataset showing new features

4. Here, we can see that we were able to break down the time series aspect and yield a little more data than we originally started with. Let's go ahead and plot the data by year:

```
plt.figure(figsize=(10,5))
ax = sns.boxplot(x='year', y='sales', data=df)
ax.set_xlabel('Year', fontsize = 16)
ax.set_ylabel('Sales', fontsize = 16)
```

After plotting our data, we will receive the following boxplot, which shows the sales for each given year. From a statistical perspective, we can confirm our initial observation that the sales are gradually increasing every year:

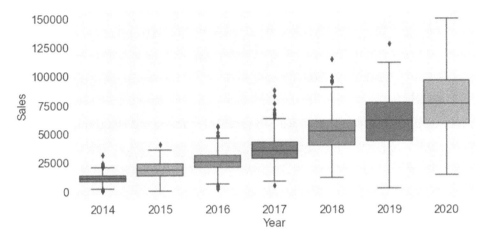

Figure 10.4 – Boxplot showing the increasing sales every year

5. Let's go ahead and plot the same graph for each given month instead:

```
plt.figure(figsize=(10,5))
ax = sns.boxplot(x='month', y='sales', data=df)
ax.set_xlabel('Month', fontsize = 16)
ax.set_ylabel('Sales', fontsize = 16)
```

Upon changing the *x*-axis from years to months, we will receive the following graph, confirming our observation that the sales data tends to peak around the January (**1**)/December (**12**) timeframes:

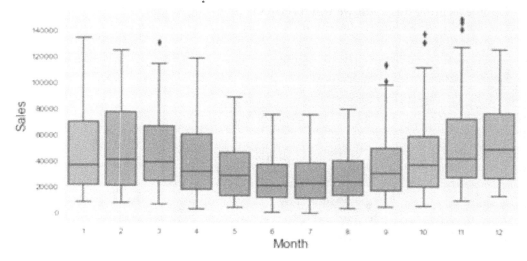

Figure 10.5 – Boxplot showing the seasonal sales for every month

Earlier, we noted that the dataset contained a great deal of noise in the sense that there was a great deal of fluctuation within the data. We can address this noise and normalize the data by taking a **rolling average** (**moving average**) – a calculation that's used to help us analyze data points by creating a series of average values.

6. We can implement this directly in our DataFrame using the `rolling()` function:

```
df["Rolling_20"] = df["sales"].rolling(window=20).mean()
df["Rolling_100"] = df["sales"].rolling(window=100).
mean()
```

7. Notice that in the preceding code, we used two examples to demonstrate the idea of a rolling average by using window values of 20 and 100. Using `Plotly Go`, we can plot the original raw data and the two rolling averages onto a single plot:

```
fig = go.Figure()
fig.add_trace(go.Scatter(x=df["Date"],
y=df["sales"], mode='lines', name='Raw Data',
line=dict(color="#bec2ed")))
fig.add_trace(go.Scatter(x=df["Date"],
y=df["Rolling_20"], mode='lines', name='Rolling 20',
line=dict(color="#858eed",dash="dash")))
fig.add_trace(go.Scatter(x=df["Date"],
y=df["Rolling_100"], mode='lines', name='Rolling 100',
line=dict(color="#d99543")))
fig.update_layout(width=800, height=500)
```

Upon executing this code, we will receive the following output:

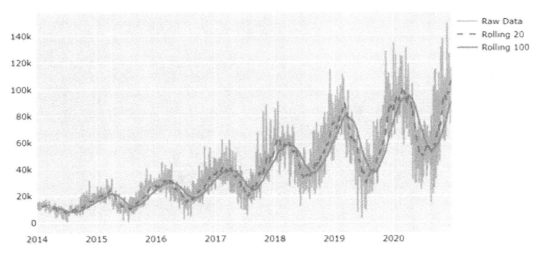

Figure 10.6 – Boxplot showing the rolling average of the sales data

Notice that the raw dataset is plotted faintly in the background, overlaid by the dashed curve representing the **rolling average** with a `window` value of 20, as well as the solid curve in the foreground representing the **rolling average** with a `window` value of 100. Using rolling averages can be useful when you're trying to visualize and understand your data, as well as building forecasting models, as we will see later in this chapter.

> **Important note**
>
> A **rolling average** (**moving average**) is a calculation that's used to smoothen out a noisy dataset by taking the moving mean throughout a particular range. The range, which is often referred to as the window, is generally the last x number of data points.

Time series data is very different from much of the datasets we have explored so far within this book. Unlike other datasets, time series data is generally thought to be consistent with several **components**, all of which we will explore in the next section.

Exploring the components of a time series dataset

In this section, we will explore the four main items that are generally regarded as the components of a time series dataset and visualize them. With that in mind, let's go ahead and get started!

Time series datasets generally consist of four main components: **level, long-term trends, seasonality,** and **irregular noise,** which we can break down into a method known as time series **decomposition.** The main purpose behind decomposition is to gain a better perspective of the dataset by thinking about the data more abstractly. We can think of time series components as being either additive or multiplicative:

$$Additive: y(t) = level + Long\ Term\ Trends + Seasonal\ Trends + \ Irregular$$

$$Multiplicative: y(t) = level * Long\ Term\ Trends * Seasonal\ Trends * \ Irregular$$

We can define each of the components as follows:

- **Level**: Average values of a dataset over time
- **Long-term Trends**: General direction of the data showing an increase or decrease
- **Seasonal Trends**: Short-term repetitive nature (days, weeks, months)
- **Irregular Trends**: The noise within the data showing random fluctuations

We can explore and visualize these compounds a little more closely using the statsmodels library in conjunction with our dataset by performing the following simple steps:

1. First, we will need to reshape our dataset by only keeping the sales column, dropping any missing values, and setting the date column as the DataFrame's **index**:

```
dftmp = pd.DataFrame({'data': df.Rolling_100.values},
                     index=df.Date)
dftmp = dftmp.dropna()
dftmp.head()
```

2. We can check the first few rows to see that the date is now our index:

	data
Date	
2014-04-10	12021.75
2014-04-11	12016.52
2014-04-12	12017.61
2014-04-13	12039.40
2014-04-14	12018.73

Figure 10.7 – First few rows of the reshaped dataset

3. Next, we will import the seasonal_decompose function from the statsmodels library and apply it to our dataframe:

```
from statsmodels.tsa.seasonal import seasonal_decompose
result = seasonal_decompose(dftmp,
model='multiplicative', period=365)
```

4. Finally, we can plot the result using the built-in plot() function and view the results:

```
result.plot()
pyplot.show()
```

Using the show() function will give us the following output:

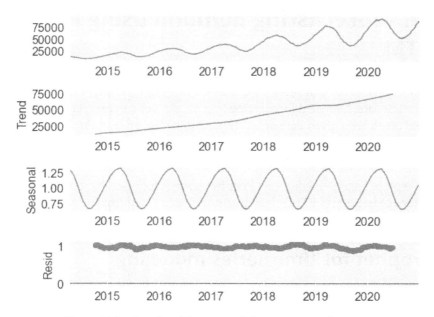

Figure 10.8 – Results of the seasonal decomposition function

Here, we can see the four components we spoke of earlier in this section. In the first plot, we can see the rolling average we calculated in the previous section. This is then followed by the **long-term trend**, which shows a steady increase throughout the dataset. We can then see the **seasonality** behind the dataset, confirming that sales tend to increase around the December and January timeframes. Finally, we can see the **residual** data or **noise** within the dataset. We can define this noise as items that did not contribute to the other main categories.

Decomposing a dataset is generally done to gain a better sense of the data and some of its main characteristics, which can often reshape how you think of the dataset and any given forecasting model that can be developed. We will learn how to develop two common forecasting models in the next section.

Tutorial – forecasting demand using Prophet and LSTM

In this tutorial, we will use the sales dataset from the previous section to develop two robust demand forecasting models. Our main objective will be to use the sales data to predict demand at a future date. **Demand forecasting** is generally done to predict the number of units to be sold on either a given date or location. Companies around the world, especially those that handle temperature-sensitive or time-sensitive medications, rely on models such as these to optimize their supply chains and ensure patient needs are met.

First, we will explore Facebook's famous **Prophet** library, followed by developing a custom **Long Short-term Memory** (**LSTM**) deep learning model. With this in mind, let's go ahead and investigate how to use the Prophet model.

Using Prophet for time series modeling

Prophet is a model that gained a great deal of traction within the data science community when it was first released in 2017. As an open source library available in both **R** and **Python**, the model was quickly adopted and widely used as one of the main forecasting models for time series data. One of the greatest benefits behind this model is also one of its consequences – its high-level nature of abstraction, allowing users to make a forecast with only a few lines of code. This limited variability can be a great way to make a quick forecast but can hinder the model development process, depending on the dataset at hand.

Over the next few pages, we will develop a Prophet model that's been fitted with our data to forecast future sales and validate the results by comparing them to the actual sales data. Let's get started:

1. To begin, let's go ahead and use the `rolling()` function to get a rolling average of our dataset. Then, we can overlay this value on the raw values:

    ```
    df["AverageSales"] = df["Sales"].rolling(window=20).
    mean()


    fig = go.Figure()
    fig.add_trace(go.Scatter(x=df["Date"],
    y=df["Sales"], mode='lines', name='Raw Data',
    line=dict(color="#bec2ed")))
    fig.add_trace(go.Scatter(x=df["Date"],
    y=df["AverageSales"], mode='lines', name='Rolling 20',
    line=dict(color="#3d43f5")))
    fig.update_layout(width=800, height=500)
    ```

2. This will result in the following output:

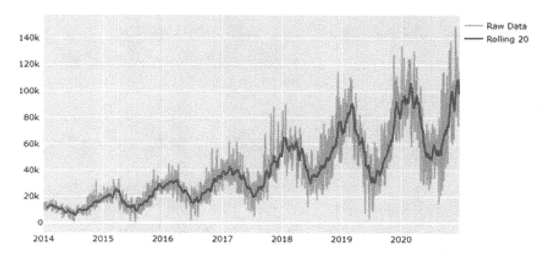

Figure 10.9 – The rolling average relative to the raw dataset

Here, we can see that the dataset is now far less **noisy** and easier to work with. We can use the **Prophet** library with our dataset to create a forecast in four basic steps:

3. First, we will need to reshape the DataFrame to integrate it with the `Prophet` library. The library expects the DataFrame to contain two columns – `ds` and `y` – in which `ds` is the date stamp and `y` is the value that we are working with. We can reshape this DataFrame into a new DataFrame using the following code:

```
df2 = df[["Date", "AverageSales"]]
df2 = df2.dropna()
df2.columns = ["ds", "y"]
```

4. Similar to the implementation of the `sklearn` library, we can create an instance of the Prophet model and `fit` that to our dataset:

```
m = Prophet()
m.fit(df2)
```

5. Next, we can call the `make_future_dataframe()` function and the number of periods of interest. This will yield a DataFrame containing a column of dates:

```
future = m.make_future_dataframe(periods=365*2)
```

6. Finally, we can use the `predict()` function to make a forecast using the future variable as an input parameter. This will return a number of different statistical values related to the dataset:

```
forecast = m.predict(future)
forecast[['ds', 'yhat', 'yhat_lower', 'yhat_upper']].
tail()
```

We can limit the scope of the dataset to a few of the columns and retrieve the following DataFrame:

	ds	yhat	yhat_lower	yhat_upper
3253	2022-12-19	135619.081957	43676.683085	224029.809289
3254	2022-12-20	135552.459718	42578.493251	225672.548118
3255	2022-12-21	135454.857473	42615.410325	225197.600026
3256	2022-12-22	135273.617248	42416.358472	225138.490115
3257	2022-12-23	135170.244729	41676.612663	226833.452043

Figure 10.10 – The output of the forecasting function from Prophet

7. Now, we can visualize our predictions using the built-in `plot()` function from our **Prophet** instance:

```
fig1 = m.plot(forecast)
```

This will result in the following output, which shows the original raw dataset, the future forecasting, as well as some upper and lower boundaries:

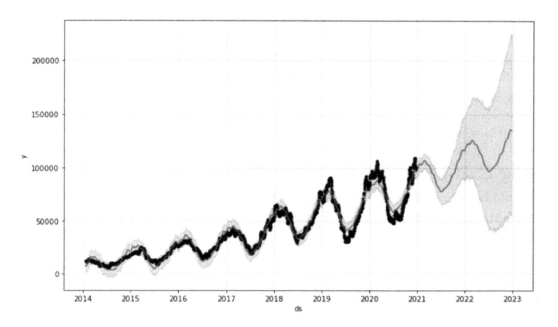

Figure 10.11 – Graphical representation of the forecasted data

8. Alternatively, we can test the model's capabilities by training the model using a portion of the data – for example, everything up to 2018. We can then use the forecasting model to predict the remaining time to compare the output with the actual data. Upon completing this, we will receive the following output:

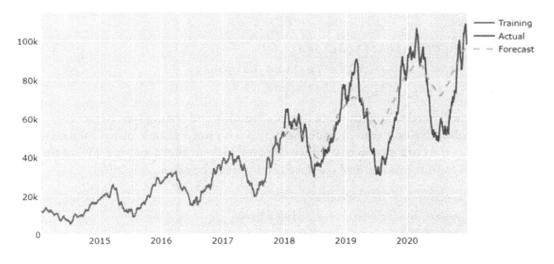

Figure 10.12 – Graphical representation of the training and testing data

Here, we can see that the dashed line, which represents the forecasted sales, was quite close to the actual values. We can also see that the model did not forecast the extremes of the curve, so it likely needs additional tuning to reach a more realistic forecast. However, the high-level nature of **Prophet** can be limiting in this area.

From this, we can see that preparing the data and implementing the model was quite fast and that we were able to complete this in only a few lines of code. In the next section, we will learn how to develop an **LSTM** using **Keras**.

Using LSTM for time series modeling

LSTM models first gained their popularity in 1997, and then again in recent years with the increase in computational capabilities. As you may recall, LSTMs are a type of **Recurrent Neural Network** (**RNN**) that can remember and forget patterns within a dataset. One of the main benefits of this model is its mid to low-level nature in the sense that more code is required for a full implementation, relative to that of **Prophet**. Users gain a great deal of control over the model development process, enabling them to cater the model to almost any type of dataset, and any type of use case. With that in mind, let's get started:

1. Using the same dataset, we can go ahead and create a rolling average using a `window` of `20` to reduce the noise in our dataset. Then, we can remove the missing values that result from this:

    ```
    df['Sales'] = df["Sales"].rolling(window=20).mean()
    df = df.dropna()
    ```

2. Using `MinMaxScaler` from the `sklearn` library, we can go ahead and scale our dataset:

    ```
    ds = df[["Sales"]].values
    scaler = MinMaxScaler(feature_range=(0, 1))
    ds = scaler.fit_transform(ds)
    ```

3. Next, we will need to split the data into our training and testing sets. Remember that our objective here is to provide the model with some sample historical data and see if we can accurately forecast future demand. Let's go ahead and use 75% of the dataset to train the model and see if we can forecast the remaining 25%:

    ```
    train_size = int(len(ds) * 0.75)
    test_size = len(ds) - train_size
    train = ds[0: train_size,:]
    test = ds[train_size : len(ds), :]
    ```

4. Given that we are working with time series data, we will need to use a `lookback` to train the model in iterations. Let's go ahead and select a `lookback` value of `100` and use our `dataset_generator` function to create our training and testing sets. We can think of a `lookback` value as the range of how far back in the data the model should look to train:

```
lookback = 100
X_train, y_train = dataset_generator(train, lookback)
X_test, y_test = dataset_generator(test, lookback)
```

5. As you may recall from our previous implementation of an LSTM model, we needed to `reshape` our data prior to using the data as input:

```
X_train = np.reshape(X_train, (X_train.shape[0], 1, X_
train.shape[1]))
X_test = np.reshape(X_test, (X_test.shape[0], 1, X_test.
shape[1]))
```

6. Finally, with the data prepared, we can go ahead and prepare the model itself. Given that we are only working with a single feature, we can keep our model relatively simple. First, we will use the `Sequential` class from Keras, and then add an LSTM layer with two nodes, followed by a `Dense` layer with a single output value:

```
model = Sequential()
model.add(LSTM(2, input_shape=(1, lookback)))
model.add(Dense(1))
```

7. Next, we can use an `Adam` optimizer with a learning rate of `0.001` and compile the model:

```
opt = tf.keras.optimizers.Adam(learning_rate=0.001)
model.compile(loss='mean_squared_error', optimizer=opt)
```

8. Recall that we can use the `summary()` function to take a look at the compiled model:

```
model.summary()
```

This will result in the following output, which provides a glimpse into the inner workings of the model:

```
Model: "sequential_2"

_____
Layer (type)                 Output Shape              Param #
=================================================================
lstm_1 (LSTM)                (None, 2)                 824

_____
dense_1 (Dense)              (None, 1)                 3
=================================================================
Total params: 827
Trainable params: 827
Non-trainable params: 0
```

Figure 10.13 – Summary of the Keras model

9. With the model compiled, we can go ahead and begin the training process. We can call the `fit()` function to fit the model on the training dataset for 10 epochs:

    ```
    history = model.fit(X_train, y_train, epochs=10, batch_
    size=1, verbose=2)
    ```

10. The model training process should be relatively quick. Once it's been completed, we can take a look at the `loss` value by visualizing the results in a graph:

    ```
    plt.figure(figsize=(10,6))
    plt.plot(history.history["loss"], linewidth=2)
    plt.title("Model Loss", fontsize=15)
    plt.xlabel("# Epochs", fontsize=15)
    plt.ylabel("Mean Squared Error", fontsize=15)
    ```

This will result in the following output, showing the progressive decrease in loss over time:

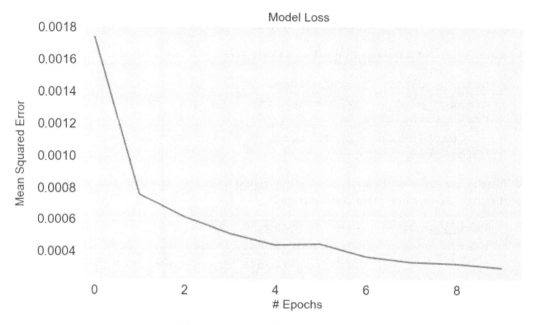

Figure 10.14 – Model loss over time

Here, we can see that the `loss` value decreases quite consistently, finally plateauing at around the 9-10 epoch marker. Notice that we specified a learning rate of `0.001` in the optimizer. Had we increased this value to 0.01, or decreased the value to 0.0001, the output of this graph would be very different. We can use the learning rate as a powerful parameter to optimize the performance of our model. Go ahead and give this a try to see what the graphical output of the loss would be.

11. With the model training complete, we can go ahead and use the model to forecast the values of interest:

```
X_train_forecast = scaler.inverse_transform(model.
predict(X_train))
y_train = scaler.inverse_transform([y_train.ravel()])
X_test_forecast = scaler.inverse_transform(model.
predict(X_test))
y_test = scaler.inverse_transform([y_test.ravel()])
```

12. With the data all set, we can visualize the data by plotting the results using `matplotlib`. First, let's plot the original dataset using `lightgrey`:

```
plt.plot(list(range(0, len(ds))), scaler.inverse_
transform(ds), label="Original", color="lightgrey")
```

13. Next, we can plot the training values using `blue`:

```
train_y_plot = X_train_forecast
train_x_plot = [i+lookback for i in list(range(0, len(X_
train_forecast)))]
plt.plot(train_x_plot, train_y_plot , label="Train",
color="blue")
```

14. Finally, we can plot the forecasted values using `darkorange` and a dashed line to distinguish it from its two counterparts:

```
test_y_plot = X_test_forecast
test_x_plot = [i+lookback*2 for i in
                list(range(len(X_train_forecast),
                len(X_train_forecast)+len(X_test_
forecast)))]
plt.plot(test_x_plot, test_y_plot , label="Forecast",
        color="darkorange", linewidth=2, linestyle="--")
plt.legend()
```

Upon executing this code, we will get the following output:

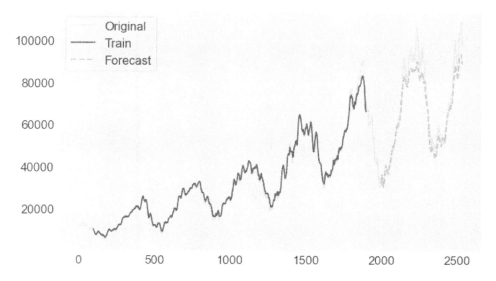

Figure 10.15 – Training and testing datasets using the LSTM model

Here, we can see that this relatively simple **LSTM** model was quite effective in making a forecast using the training dataset we provided. The model was not only able to capture the general direction of the values, but also managed to capture the seasonality of the values as well.

Summary

In this chapter, we attempted to analyze and understand time series data, as well as developing two predictive forecasting models in less than 15 pages. We began our journey by exploring and decomposing time series data into smaller features that can generally be used with shallow machine learning models. We then investigated the components of a time series dataset to understand the underlying makeup. Finally, we developed two of the most common forecasting models that are used in the industry – the Prophet model, by Facebook, and an LSTM model using Keras.

Throughout the last few chapters, we have developed various technical solutions to solve common business problems. In the next chapter, we will explore the first step in making models such as these available to end users using the Flask framework.

Section 3: Deploying Models to Users

Thus far, we have discussed Python, using data, and developing models. In this section, we will explore how these models can be moved to production and made available to an end user. We will explore four commonly used platforms for deploying a machine learning model: AWS, Heroku, GCP, and Python anywhere.

This section comprises the following chapters:

- *Chapter 11, Deploying Models with Flask Applications*
- *Chapter 12, Deploying Applications to the Cloud*

11
Deploying Models with Flask Applications

Over the course of this book, we explored the development of numerous robust machine learning models in areas such as breast cancer detection, scientific topic modeling, protein classification, and molecular property prediction. In each of these tutorials, we prepared and validated our models to allow them to have the best predictive power possible. We will now pivot from the development of new models to the deployment of trained models to our end users.

Within this chapter, we will explore one of the most popular frameworks for the preparation of web applications: **Flask**. We will use Flask to prepare a web application to serve our models to end users, and we will also prepare an **Application Programming Interface** (**API**) to serve our predictions to other web applications.

Over the course of this chapter, we will cover the following topics:

- Understanding API frameworks
- Working with Flask and Visual Studio Code
- Using Flask as an API and web application
- Tutorial – Deploying a pretrained model using Flask

With these objectives in mind, let's go ahead and get started!

Understanding API frameworks

Whether you are logging in to your email account, scrolling through social media, or even logging in to an online retailer, we use **web applications** on a daily basis to accomplish a variety of tasks. For example, imagine a user scrolling through an electronic laboratory notebook on their local computer. When the user logs in and sees their data, this information is retrieved using an API (that is, an *application programming interface*, not to be confused with an *active pharmaceutical ingredient*). Once the data is retrieved for the user in the backend, it populates the frontend in a beautiful **User Interface** (**UI**) that allows the user to interact with the data, make changes, and save it. We can use web applications and APIs in a variety of ways, such as transferring data, communicating with others, or even making predictions, as illustrated in *Figure 11.1*:

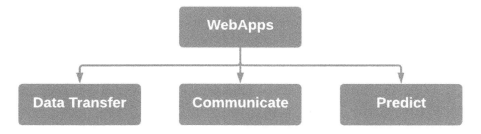

Figure 11.1 – Some examples of web application functionality

With all of these capabilities, APIs and their counterparts have provided the main tool in the web application space for creating UIs to serve data and make predictions. There are a number of useful **web application frameworks** available for a range of programming languages, as illustrated in *Figure 11.2*:

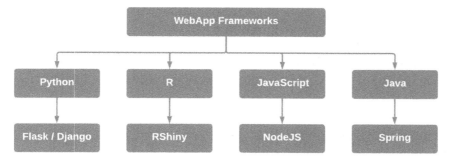

Figure 11.2 – Some examples of web application frameworks

For the purposes of this chapter, we will focus on one of the more popular machine learning deployment frameworks: Flask (`https://github.com/pallets/flask`). Relative to its counterparts, Flask can be thought of as a **micro web framework** – it is completely written in Python and highly abstracted, allowing users to get started in the model deployment process with little to no difficulties.

As we begin to deploy models using the Flask framework, it is important to ask ourselves who the end user of our application will be. In many cases, predictions using our previously trained models will be conducted by colleagues and stakeholders. Therefore, having a useable UI will be important. On the other hand, our deployed models may not be needed by a person but rather a piece of software or another web application that will need to programmatically interact with it. In that case, a UI will not be needed – however, we will need an organized way (for example, **JSON**) to handle the transfer of data between the two systems. We can see a depiction of these two cases in *Figure 11.3*:

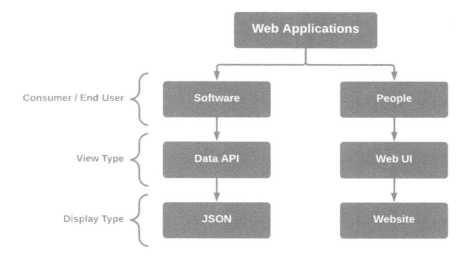

Figure 11.3 – The two general types of web applications

In either case, we will be able to accommodate both of these cases using Flask. The Flask framework offers a variety of architectures – both simple and complex – allowing users to select the pattern that best fits their needs. Flask APIs, in a similar way to their counterparts such as **Django**, **Node.js**, and **Spring**, all generally operate in a similar manner using URLs. For both backend APIs and frontend UIs, we can use URLs to organize how we develop an application. For example, users can log in to a website to view and edit data within their profiles, whereas APIs can allow external entities to interact with models, as depicted in *Figure 11.4*:

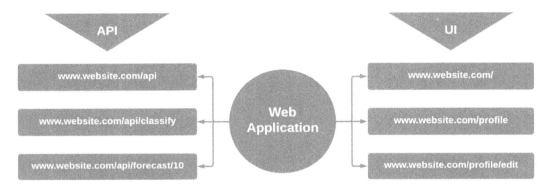

Figure 11.4 – The two general types of web applications with examples

In order to interact with a web application, a user needs to make what is known as an **HTTP request**, which is usually carried out without them knowing. Each of these requests is generally associated with a URL, allowing the user to accomplish a task. The four HTTP request types are depicted in *Figure 11.5*:

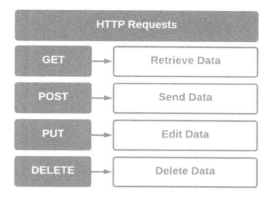

Figure 11.5 – The four HTTP request types

For example, if a user navigating to www.website.com/profile intends to retrieve the details of their profile, they would use a GET request. On the other hand, an application using the API with the intention of classifying a segment of text would use a POST request to send the text to www.website.com/api/classify. These URL paths are known as *routes* within the confines of web applications, and they allow developers and data scientists to better organize their models for deployment. In the following section, we will see how routes can be used more specifically within the Flask framework.

Working with Flask and Visual Studio Code

Flask is one of the most commonly used and versatile web applications available in the **Python** language. Its abstract and **high-level framework** makes it easy for users of all levels to have an implementation up and running in no time. Over the course of this section, we will learn about the different components of a Flask application and deploy a simple model locally on our machine.

Before we can get started with Flask, we will need an **Integrated Development Environment (IDE)** to work with. So far, we have worked almost exclusively in **Jupyter Notebook** to train and develop models. When it comes to implementation, we will need another type of IDE to work with. There are numerous Python IDEs we can use, such as **PyCharm**, **Spyder**, or **Visual Studio Code** (**VSC**). I personally have found VSC to be the most user-friendly to work with, and therefore, we will use that as our primary IDE in this section. You can download VSC from their website (https://code.visualstudio.com/download) or by using **Anaconda**.

344 Deploying Models with Flask Applications

Go ahead and begin the installation process, which might take a few minutes. While you wait, create a new folder called `flask-test` on your local computer. Once the installation process is complete, open VSC. You can open the folder you just created in a few simple steps:

1. Click **File** on the top menu.
2. Click **Open Folder**.
3. Navigate to your directory and click **Select Folder**.

You should now see the name of your directory in the **explorer** pane on the left-hand side of the screen. Within the explorer pane, you will be able to see all of the files and folders relevant to your current project. Let's go ahead and populate it with a file called `app.py` by right-clicking in the explorer pane and selecting **New File**.

The `app.py` file is the main file that Flask uses in its framework. Everything within the application is included in this file or referenced from within it. Although its content depends on the exact implementation of the user, the file generally contains four main sections:

1. Importing libraries, data, and other resources
2. Instantiating the application and declaring other useful functions
3. Declaring the routes for the application
4. Running the `__name__ == "__main__"` driver piece of code

We can see an illustration of these components in *Figure 11.6*:

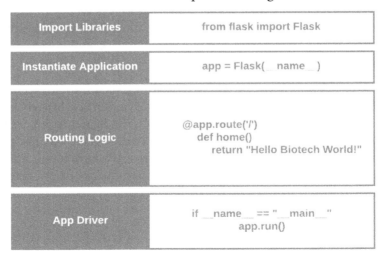

Figure 11.6 – The main components of a Flask application

Let's now go ahead and populate app.py with some code. This is generally done in four main sections:

1. We will begin by importing the Flask class from the flask library:

    ```
    from flask import Flask
    ```

2. Next, we will need to create an instance of our Flask app:

    ```
    app = Flask(__name__)
    ```

3. We can now use the app object to create routes for our application. Routes operate by executing the function directly beneath it when that route is interacted with. Let's make a simple one that returns "Hello Biotech World!":

    ```
    @app.route('/')
    def biotech():
        return "Hello Biotech World!"
    ```

4. Finally, we will need a driver for the application that can fulfill using if __name__ == '__main__'. We will also set the debug parameter as True to help us address any potential issues, and we will set the port value to 8080:

    ```
    if __name__ == '__main__':
        app.run(debug=True, port=8080)
    ```

From the command line in VSC, go ahead and run the Python application:

```
$ python3.8 app.py
```

This will run the application on your local computer. You can access it through any browser, such as **Google Chrome**, and by navigating to http://localhost:8080/. Upon reaching this URL, you should be greeted by our previous message. Please note that the localhost URL is a link only accessible locally on your computer and is not available to others. The concept of routes should be familiar to us from the many websites we have used in the past. We can break down a URL into its smaller components, as depicted in *Figure 11.7*:

Figure 11.7 – The main components of a URL

In our case, we are currently editing the path or endpoint of the application. Flask applications can handle many paths and endpoints, giving developers a great deal of flexibility.

You can stop the application from running by pressing *CTRL + C* in the command line, which will halt the process. With the process halted, go ahead and create a second route by copying the current route and function directly below it. Give the path a value of `/lifescience` (instead of just `/`) and give its function a unique name such as `lifescience`. Next, change the returned value, run the application again, and navigate to `http://localhost:8080/lifescience`. If all was successful, you should be able to see your new message!

> **Routes and Functions**
>
> Please note that routes must be unique – this means that you cannot have multiple routes in Flask pointing to `/biotech`. Similarly, the function beneath the route must also be unique in its name.

When deploying our models, we will work with similar architecture. However, the return statements will generally comprise either a UI for people to use or data for applications to consume. In the following section, we will explore this in a little more depth by using a **Natural Language Processing** (**NLP**) use case.

Using Flask as an API and web application

In *Chapter 9*, *Natural Language Processing*, we explored the use of the `transformers` library for the purposes of running text similarity search engines. By using this technology, we could have explored other models and implementations, such as **sentiment analysis**, **text classification**, and many more. One particular type of model that has gained a great deal of traction when it comes to NLP is the **summarization** model.

We can think of summarization models as tasks designed to reduce several paragraphs of text down to a few sentences, thereby allowing users to reduce the amount of time required to read. Luckily for us, we can implement an out-of-the-box summarization model using the `transformers` library and install that in our `app.py` file. Not only will we need to cater to human users (by using a UI), but we will also need to cater to web applications (APIs) that may be interested in using our model. In order to accommodate these two cases, we will need three files in total within our project to get us started:

- `app.py`: This is the main file in which the Flask framework and all NLP models are instantiated.
- `styles.css`: This is a CSS file that allows us to style the UI.

- `index.html`: This is an HTML file with a pre-built UI page that human users will interact with.

For better organization, let's add the CSS file to a directory called `styles` and the HTML file to a directory called `templates`.

When working with new Flask applications, we generally want to have a *blank slate* when it comes to the libraries we installed via `pip`. In other words, each Flask application should have its own *virtual environment*, where we only install libraries the application will need and use. We can accomplish this using `virtualenv`, which (ironically) can be installed using `pip`.

Once installed, we can use `virtualenv` on the command line to create a new environment for this project called `.venv`:

```
$ python38 virtualenv .venv
```

You can call your virtual environment anything you like, but most users generally default to the name in the preceding command. You will know this command was successful when you see a new directory in your current working directory with the specified name. We will now need to *activate* the environment, which can be a little tricky depending on the type of system you are using. **Windows** users can activate their environment using the following command:

```
> .\.venv\Scripts\activate
```

On the other hand, **Linux** and **Mac** users can activate their environments via the following command:

```
$ source .venv\bin\activate
```

You can confirm the environment was activated if its name appears on the left-hand side of the command line's current working directory. Go ahead and install `flask` and `transformers`, as we will need these libraries in the current environment.

With the environment set up and including the three files discussed, we should have a directory structure as depicted in *Figure 11.8*:

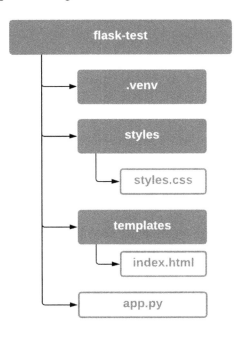

Figure 11.8 – The current folder structure of this project in VSC

With the project structure now in place, let's add some code to app.py. We can begin by importing some of the libraries we will need within this application:

```
from flask import Flask, jsonify, request, render_template
import json
from transformers import pipeline
import re
```

Now that the libraries have been imported, we can instantiate an instance of the Flask application just as before. However, we will need to specify the template folder this time:

```
app = Flask(__name__, template_folder='templates')
```

With the application instantiated, we can now create an instance of the summarizer model from the transformers pipeline class:

```
summarizer = pipeline("summarization")
```

Next, we can add our *routes*. We will first create a route to our home page, which displays the UI using the `index.html` file:

```
@app.route('/')
def home():
    return render_template('index.html')
```

We will then need to add two routes: one for the UI, and another for the API. There are many best practices that vary depending on the framework, the industry, and the use case. In most scenarios, `api` endpoints are generally preceded with the `api` word to distinguish them from others. Let's go ahead and create a route for the UI first:

```
@app.route('/prediction', methods = ["POST"])
def ui_prediction():
    """
    A function that takes a JSON with two fields: "text" &
"maxlen"
    Returns: the summarized text of the paragraphs.
    """
    print(request.form.values())
    paragraphs = request.form.get("paragraphs")
    paragraphs = re.sub("\d+", "", paragraphs)
    maxlen = int(request.form.get("maxlen"))
    summary = summarizer(paragraphs, max_length=maxlen, min_
length=49, do_sample=False)
    return render_template('index.html', prediction_text = '"
{} "'.format(summary[0]["summary_text"])), 200
```

Notice that within this function, we use the `request.form.get` function to retrieve the values from the form in the UI. In addition, we use some regular expressions to clean up the text, and then we summarize the contents using the summarizer model. Finally, we return the summary and the `index.html` file.

Let's now create the second route for the api:

```
@app.route('/api/prediction', methods = ["POST"])
def api_prediction():
    """
    A function that takes a JSON with two fields: "text" &
"maxlen"
    Returns: the summarized text of the paragraphs.
```

```
"""
    query = json.loads(request.data)
    paragraphs = re.sub("\d+", "", query["text"])
    maxlen = query["maxlen"]
    minlen = query["minlen"]
    summary = summarizer(paragraphs, max_length=maxlen, min_
length=minlen, do_sample=False)
    return jsonify(summary), 200
```

Notice that in addition to taking the input data, cleaning the contents, and summarizing it, we can take the `maxlen` and `minlen` parameters directly from the JSON object.

Finally, we can go ahead and execute the code:

```
if __name__ == '__main__':
    app.run(debug=True)
```

With that, we have successfully developed the Flask application. Once deployed, you should be able to navigate to `http://localhost:5000/` and start summarizing paragraphs of text! We can see an example of the application in *Figure 11.9*:

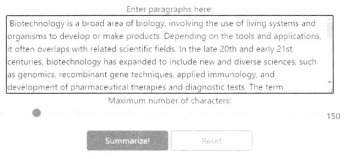

Figure 11.9 – A screenshot of the summarizer web application

In addition, we can use applications such as **Postman** (https://www.postman.com/) to test the API endpoint. Alternatively, we could use the requests library from Python to accomplish the same thing. In this case, we would need to make a POST request, add the URL, and then add the data in the form of a dictionary and the content type of the application/JSON:

```
{
    "text" : "Biotechnology is a broad area of biology,
involving the use of living systems and organisms to develop or
make products.

                                ...

 molecular biology, biochemistry, cell biology, embryology,
genetics, microbiology) and conversely provides methods to
support and perform basic research in biology.",
    "maxlen" : 60,
    "minlen" : 30
}
```

With the application now working, we managed to successfully create a solution that uses Flask to cater to both human users and other web applications. In the final chapter of this book, we will deploy this application to the cloud. However, one of the most important steps of doing this is providing a list of the libraries that need to be installed. Given that we have set up a virtual environment, we can easily transfer a list of these libraries to a requirements.txt file via pip:

```
$ pip freeze > requirements.txt
```

With that, you should now see a requirements.txt file in the same directory as app.py. It is important to ensure that the environment you use only contains the libraries you plan to use. This helps keep the application light and fast to use. In the following section, we will look at a more in-depth application – one that uses a previously trained model concerning the breast cancer dataset we saw earlier in this book.

Tutorial – Deploying a pretrained model using Flask

In the previous example of creating a Flask application, we saw how we can make use of the application in conjunction with a predictive model to deploy a solution to our end users. However, the model that we deployed was an out-of-the-box solution and not a model we developed ourselves. In this section, we will once again deploy a model within a Flask application; however, we use a model based on the cancer dataset we saw in *Chapter 5, Understanding Machine Learning*.

If you recall, the main idea behind this model was to take in a number of measurements for a given tumor, and based on those measurements, determine what the diagnosis will likely be, resulting in either `Malignant` or `Benign`. Within this application, we will enable users to interact with a trained model and enter measurements that the model will use to make a prediction. With this in mind, let's get started!

In the same way as before, go ahead and add a new folder and a new virtual environment to install the relevant libraries.

Using the same directory architecture and process as before, we can begin by importing the relevant libraries. Notice that we have added the `pickle` library here, as we will need to use the *pickled* models we previously created:

```
from flask import Flask, jsonify, request, render_template
import json
import pickle
import pandas as pd
from sklearn.preprocessing import StandardScaler
```

Our next step involves importing the two models we trained – the actual classification model and the standard scaler model we used for the data:

```
loaded_scaler= pickle.load(open("./models/ch10_scaler.
pickle",'rb'))
loaded_clf= pickle.load(open("./models/ch10_rfc_clf.
pickle",'rb'))
```

We can then define a `predict_diagnosis` function to clean up our code later when developing our routes. This function will take the input data in the form of a list, the scaler model, and the classification model:

```python
def predict_diagnosis(inputData, scaler, model):
    """
    Function that takes a list of measurements, scales them,
and returns a prediction
    """
    inputDataDF = pd.DataFrame([inputData])
    scaledInputData = scaler.transform(inputDataDF)
    prediction = model.predict(scaledInputData)
    return prediction[0]
```

Next, we will instantiate the Flask application while specifying the `template` folder:

```python
app = Flask(__name__, template_folder='templates')
```

With these items taken care of, we can focus on our routes. First, we will create a home route that users will see first:

```python
@app.route('/')
def home():
    return render_template('index.html')
```

Next, we will need a `prediction` route, just as before. The only difference here is that the number of input values will be greater, as we are working with a few more features now:

```python
@app.route('/prediction', methods = ["POST"])
def prediction():
    print(request.form.values())
    radius_mean = request.form.get("radius_mean")
    texture_mean = request.form.get("texture_mean")
    smoothness_mean = request.form.get("smoothness_mean")
    texture_se = request.form.get("texture_se")
    smoothness_se = request.form.get("smoothness_se")
    symmetry_se = request.form.get("symmetry_se")
    input_features = [radius_mean, texture_mean, smoothness_
mean, texture_se, smoothness_se, symmetry_se]
    prediction = predict_diagnosis(input_features, loaded_
```

```
scaler, loaded_clf)
    prediction = "Malignant" if prediction == "M" else "Benign"


    return render_template('index.html', prediction_text = '"
{} "'.format(prediction))
```

Finally, we can go ahead and run the application:

```
if __name__ == '__main__':
    app.run(debug=False, port=5000)
```

Upon running the model and navigating to localhost in the web browser, the application will appear. Go ahead and try making a few predictions using the UI, an example of which is displayed in *Figure 11.10*:

Figure 11.10 – A screenshot of the breast cancer web application

We can see that the model is able to take our input data, run a prediction, and return a result to the user. One thing we did not do here is create an API route for other web applications to interact with our model. As a challenge, go ahead and create this route, using the previous summarization application as an example.

Summary

In this chapter, we steered away from the development of models and focused more on how models can be deployed to interact with web applications. We investigated the idea of data transfer via APIs, and we also learned about some of the most common frameworks. We investigated one of the most common Python web application frameworks known as Flask. Using Flask, we developed an NLP summarization model that allows both human users and other web applications to interact with it and use its capabilities. In addition, we learned how to deploy previously trained models, such as those from `scikit-learn`.

In each of these instances, we launched our models locally as we developed their frameworks and capabilities. In the next chapter, we will make our model available to others by using **Docker** containers and **AWS** to deploy our model to the cloud.

12
Deploying Applications to the Cloud

In the previous chapter, we focused our efforts on integrating our models within the **Flask** framework to develop two main methods of serving data to end users: **Graphical User Interfaces (GUIs)**, and **Application Programming Interfaces (APIs)**. Using the Flask framework, we managed to locally deploy our models for development purposes only. In this chapter, we will take the next step forward and deploy our model to the cloud, thus making it available not only locally to ourselves but also across the web to many other users.

There are many different deployment platforms out there, such as **Amazon Web Services (AWS)**, **Google Cloud Platform (GCP)**, Azure, and Heroku, each of which serves to fulfill a number of needs. In each of these platforms, there are a number of solutions, each containing its respective pros and cons. For each of these solutions, there are a number of ways we can deploy a framework within them. Essentially, the number of possible ways to deploy a solution is practically uncountable, and because of this, users can get easily overwhelmed. Over the course of this chapter, we will explore some of the most common and straightforward paths new developers generally take to deploy their applications.

Throughout the following sections, we will cover the following topics:

- Exploring current cloud computing platforms
- Understanding containers and images
- Tutorial – deploying a container to AWS (Lightsail)
- Tutorial – deploying an application to GCP (App Engine)
- Tutorial – deploying an application's code to GitHub

With these objectives in mind, let's go ahead and get started!

Exploring current cloud computing platforms

One of the most significant technology trends over the last few years has been the shift to cloud computing. Although most companies used to prefer to own, operate, and maintain their own data centers and infrastructure, most enterprises around the globe now maintain a cloud-first approach. There are many reasons as to why companies have moved down this path, such as reduced costs, scalability, security, and much more. Given the surge in demand for cloud computing capabilities, a number of cloud computing platforms began to grow and expand in the market in response to this major movement in the digital world.

Over the last few years, many of these cloud computing platforms began to not only develop solutions to meet major infrastructure needs but also focus on targeted needs specifically within the field of data science. The main platforms are **AWS**, **GCP**, and **Microsoft Azure**, as depicted in the following diagram:

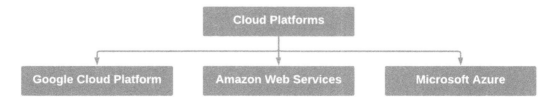

Figure 12.1 – Some common cloud computing platforms

Many companies around the world generally operate at the **enterprise** level with one of these providers for consistency. From the perspective of data scientists and developers, these platforms are nearly identical as they generally contain very similar solutions to meet our needs.

Each of these platforms includes numerous solutions designed to deploy frameworks and make them available to end users in some form. The degree to which these resources are provided to end users is generally the difference between these platforms. For example, a developer may expect high levels of activity for a given web application and may therefore decide to deploy their model using **AWS Elastic Beanstalk** or **Amazon Elastic Container Service (ECS)**. On the other hand, another user may only wish to deploy their web application to a few users and in the simplest way possible and would therefore choose to use **Amazon Elastic Compute Cloud (EC2)**. In any case, the specific solution a developer chooses is generally selected based on the specific need. Let's go ahead and take a look at some of the most popular solutions out there, as depicted in the following screenshot:

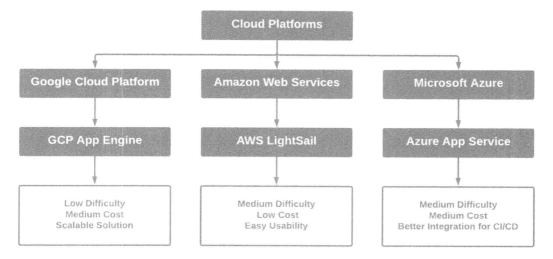

Figure 12.2 – Some of the most common tools to deploy web applications

When it comes to deploying **Flask** applications, three deployment solutions have gained a great deal of traction in recent years within the data science community. Each of these solutions contains its respective pros and cons, and it is the responsibility of the data scientist or developer to ensure that the business needs for a given web application are matched to the best possible solution in any of these platforms.

As web applications began to increase in popularity, it soon became evident that consistency within the applications was needed to ensure that an application that a developer creates and deploys within one platform can be just as easily deployed in another platform with the smallest number of changes. In the following section, we will discuss the idea of containerization using an example known as **Docker**.

Understanding containers and images

One of the easiest ways to build, deploy, and manage a web application is through the use of **containers**. We can think of containers as buckets or vessels containing all items that make up a web application but in the form of an **Operating System (OS) virtualization**. Think back for a moment to the previous chapter—*Chapter 11, Deploying Models with Flask*—in which we created a virtual environment to better maintain the packages we needed to install for the application. Containers can be thought of in quite a similar way, only on the OS level.

Containers consist of a number of items such as executables, libraries, binary code, and much more. Given that they do not contain some of the heavier items servers tend to have such as OS images, they are considered to have less overhead, making them more lightweight. Since these lightweight containers are considered to be packaged up and ready to go, developers (or automated systems) are able to easily deploy multiple instances of these containers to meet the needs of the increased traffic to a given website or application.

Understanding the benefits of containers

There are many benefits to using a container when managing and deploying web applications, especially at the enterprise scale. Ultimately, they provide a consistent and streamlined way to build, deploy, and manage multiple applications in an efficient manner. A few of the main benefits include the following:

- **Greater scalability**—Easily deploy more instances to meet a given need.

- **Increased portability**—Deploy to different platforms and OSes.

- **Reduced overhead**—Use fewer resources than traditional methods.

Containers are highly effective in many different areas—two in particular fall within the scope of microservices and automation. In the case of microservices, applications are generally broken down into smaller components in which each component needs to be deployed and scaled independently of the others. It is no surprise that containers in this case would be an excellent solution for the given problem. On the other hand, in the case of automation, containers can be easily created or removed in an automated fashion, making them very useful for scalability, as well as for **Continuous Integration/ Continuous Deployment (CI/CD)** pipelines.

> **Important note**
> We can think of CI/CD pipelines as methods to automate software delivery in order to standardize processes and reduce human error. There are generally four stages for any given CI/CD pipeline: new code being pushed to a repository such as GitHub (which we will see later in this chapter), a building script that builds or compiles the code, a testing script that tests certain parts of the code, and finally the deployment platform that hosts the final product.

In the following tutorial, we will explore the process of deploying a container—specifically, a Docker container—to AWS Lightsail.

Tutorial – deploying a container to AWS (Lightsail)

AWS Lightsail is a managed cloud platform that has gained a great deal of popularity in recent years due to its simple interface and fast deployment capabilities and is overall a great way to get started when deploying applications using AWS. Some of the most common use cases for using Lightsail as opposed to other AWS products or solutions include simple **Machine Learning** (**ML**) web applications (such as ours!), static portfolio websites, and dynamic e-commerce websites, as well as simple APIs.

Over the course of this tutorial, we will deploy our Flask application to AWS Lightsail using the AWS **Command Line Interface** (**CLI**). You can install the CLI by navigating to the AWS CLI page (https://docs.aws.amazon.com/cli/latest/userguide/install-cliv2.html), selecting an OS of interest, and following the installation instructions. You can confirm that the CLI was correctly installed by running the following command:

```
$ aws configure --profile
```

If the installation was properly completed, you will be guided through the configuration process. Please go ahead and configure the CLI as needed. The process may request **Identity and Access Management** (**IAM**) credentials—go ahead and provide it with the proper credentials from our previous examples or prepare a new set of credentials, as we have previously done in *Chapter 9*, *Natural Language Processing*, specifically in the *Working with unstructured data* section.

With the CLI and credentials all set, let's now once again focus on the application, starting with the content. If you recall from *Chapter 11*, *Deploying Models with Flask Applications*, the content of the application should now include the virtual environment, `styles.css`, `index.html`, and, of course, `app.py`, as depicted in the following screenshot:

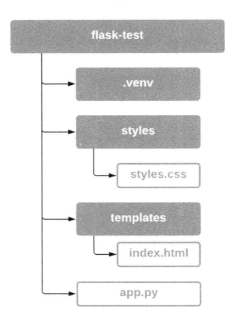

Figure 12.3 – The content of our current working directory

In order to deploy our application, we will need a few more files that we will soon explore to help both with the deployment process and the containerization process. In addition to AWS Lightsail, we will also be using **Docker**—a commonly used tool that allows users to create, deploy, and run applications in isolated containers. You can download Docker for your specific OS by visiting the Docker website (https://docs.docker.com/get-docker/). Let's go ahead and explore these new files, their content, and how they will be used within the application.

First, we begin with the Dockerfile, which contains a set of instructions to prepare the environment for the application. These instructions include the version of Python, establishing the working directory, copying the files of interest, and of course, installing the requirements. The code is illustrated in the following snippet:

```
FROM python:3.8
EXPOSE 5000/tcp

WORKDIR /app
```

```
COPY requirements.txt .
COPY models/ch10_scaler.pickle /models/ch10_scaler.pickle
COPY models/ch10_scaler.pickle /models/ch10_rfc_clf.pickle
COPY styles /app/styles
COPY models /app/models
COPY templates /app/templates
COPY app.py .

ENV IN_DOCKER_CONTAINER Yes

RUN pip install --upgrade pip
RUN pip3 install -r requirements.txt

CMD [ "python", "./app.py" ]
```

With the Dockerfile prepared, we can now go ahead and build the container image. We can build the container using Docker, which we installed earlier, by executing the following command:

```
$ docker build -t flask-container .
```

Upon executing this command (don't forget the . at the end, which signifies the current directory!), Docker will build a container tagged as flask-container.

Our next step will be to create a container service using the AWS CLI. We can do this by executing the following code, in which we specify service-name, power, and scale parameters. Note that these parameters specify the capacity of the service:

```
$ aws lightsail create-container-service --service-name flask-
service --power small --scale 1
```

Upon executing this command, you should be able to monitor the progress. Once the service changes from a state of pending to active, you can execute the next command, which pushes the container image:

```
$ aws lightsail push-container-image --service-name flask-
service
--label flask-container --image flask-container
```

Upon executing this command, you will see the following value in the results:

```
":flask-service.flask-container.X"
```

Please note that X should be a numeric value that corresponds to the time the image was pushed to the container service. If this is the first time you have done this, the value should be 1.

Next, we will need to create a file called `containers.json`, specifying the Flask image as well as the port, containing the following code:

```
{
    "flask": {
        "image": ":flask-service.flask-container.X",
        "ports": {
            "5000": "HTTP"
        }
    }
}
```

Go ahead and replace X with the numerical value you received before. Again, if this is your first time deploying the container, the value should be 1. With that done, we can now go ahead and create our final file called `public-endpoint.json`, which specifies the container name and port, containing the following code:

```
{
    "containerName": "flask",
    "containerPort": 5000
}
```

At this point, the hierarchy of the directory should include all of the previous files in addition to `containers.json`, `Dockerfile`, and `public-endpoint.json`, as depicted in the following figure:

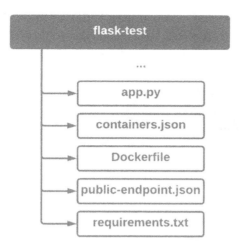

Figure 12.4 – The content of our current working directory

Now that the files and containers are all in order, we can now proceed with the final steps of deploying the container to end users. In order to do so, we can use the `create-container-service-deployment` command with the following code:

```
$ aws lightsail create-container-service-deployment --service-
name flask-service --containers file://containers.json
--public-endpoint file://public-endpoint.json
```

Upon executing the code, you should see the state of the application be listed as **Deploying**. Once the state changes to **Running**, you can use the `get-container-services` command to monitor the current application by executing the following command:

```
$ aws lightsail get-container-services --service-name flask-
service
```

Upon completion of the command, you will see a **Uniform Resource Locator** (**URL**) as output. Go ahead and navigate to the URL listed, and you should be able to see the application we developed online and available to our end users. We can see an example of this in the following screenshot:

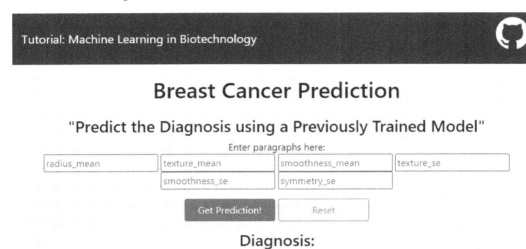

Figure 12.5 – Web application running on AWS Lightsail

Alternatively, you may want to view the application using the management console on AWS. To do so, navigate the console and search for AWS Lightsail. You should be redirected to the AWS Lightsail page where you should see your instance and container, as depicted in the following screenshot:

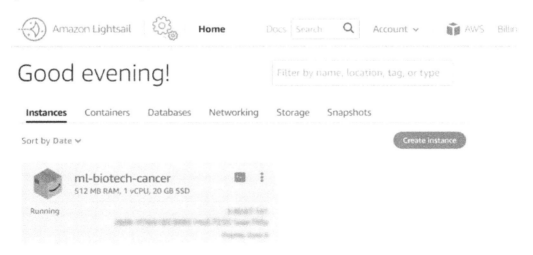

Figure 12.6 – AWS Lightsail management console

In this tutorial, we successfully took our local implementation of Flask and deployed it as a web application to AWS. In the following tutorial, we will deploy the same application to GCP's App Engine.

Tutorial – deploying an application to GCP (App Engine)

Over the course of this tutorial, we will deploy the same application to GCP's App Engine. One of the biggest benefits of GCP relative to most other cloud platforms is its ease of use while ensuring users can deploy models with minimal problems and errors. With that in mind, let's go ahead and deploy our application to GCP.

We can begin by installing the **GCP CLI**. Go ahead and navigate to the installation page (`https://cloud.google.com/sdk/docs/install`), select the OS that best corresponds to your machine, and follow the installation instructions. Once completed, you will be asked if you would like to log in to GCP. Go ahead and answer yes using the `y` key, as shown here:

```
You must log in to continue. Would you like to log in (Y/n)?   y
```

You will be redirected to your browser where you can log in using your Google account. Go ahead and log in using the same Google credentials you used in our previous adventure in *Chapter 7, Supervised Machine Learning*, concerning GCP.

Once logged in, you will be prompted to select a project. Select the project you previously created earlier in this book. Go ahead and complete any other remaining items such as the default region, and finish up the configuration.

Once completed, you should have the `gcloud` CLI installed on your system. Be sure to restart the command-line window you are using, as some `PATH` variables may need to be refreshed.

With the CLI now installed and working, we can go ahead and get started. Navigate to the `flask_cancer_ae` directory found in the accompanying code. We will need to create a new file called `app.yaml` in our directory, containing the following code:

```
runtime: python37
```

This will simply specify the runtime for our application. With this file saved, we can go ahead and do some preliminary configuration. We will first need to set the project **Identifier** (**ID**), if we've not already done so, using the following command:

```
$ gcloud config set project GCP-PROJECT-ID
```

Be sure to replace GCP-PROJECT-ID with your associated project ID. With that set, we will now need to enable the **Cloud Build CLI**, which is used to create a container for the application using our files, with the following command:

```
$ gcloud services enable cloudbuild.googleapis.com
```

Next, we will initialize the application in **App Engine** for this particular project. We can do so using the following command:

```
$ gcloud app create --project= GCP-PROJECT-ID
```

Be sure to replace GCP-PROJECT-ID with your particular project ID. Finally, in order to go ahead and deploy the project, we can use the following command:

```
$ gcloud app deploy
```

Once the process is complete, the project will be deployed to GCP! We can check the application using the browse command, as shown here:

```
$ gcloud app browse
```

In addition to using the CLI, we can also visit the **App Engine dashboard** found in the GCP console to complete a number of tasks, such as the following:

- Access the application.
- Monitor traffic.
- Review billing.
- ...and much more!

GCP has many wonderful capabilities, giving users a great experience when it comes to deploying applications, managing data, and monitoring traffic. If you are interested in learning more about GCP, I highly encourage you to follow and complete the many great tutorials provided by the GCP platform.

With our application now deployed to GCP, our next step will be to explore a different way to send our code elsewhere: through the git CLI. In the following section, we will explore the process of pushing code to GitHub.

Tutorial – deploying an application's code to GitHub

Over the last two tutorials, we deployed our applications to cloud platforms in order to allow users to interact with our models using the Flask framework. In the first platform, we used AWS Lightsail, and in the second, we used GCP's App Engine. In this tutorial, our objective will be to deploy our code, not with the intent of making the models available to users but to showcase our code and hard work to other data scientists, as well as potential future employers. We will do so by deploying our code using **GitHub**.

All of the coding examples and tutorials throughout this book have been made available online using GitHub. If you have not done so already, I highly encourage you to create your own account. You can think of GitHub as LinkedIn for coders—a space to showcase your hard work to others.

You can create a free GitHub account by navigating to their main website (`https://github.com/`) and registering as a new user. Once you are registered, you will be able to save your code and work with projects or repositories. You can think of a repository as a space to save your work, with multiple versions being saved so that users can revert to older code when needed.

The way this works is that users will have an instance or copy of a given project or repository locally on their computers. Every now and then, when significant progress has been made, a user can make an update or commit, and then push those new changes to the remote repository for backup, as depicted in the following figure:

Remote Repo   Local Repo
(www.github.com)   (Local)

Figure 12.7 – Local and remote repositories

With your profile created, let's go ahead and install `git` on the command line. We can get things going by navigating to `https://git-scm.com/book/en/v2/Getting-Started-Installing-Git` and installing `git` for your given OS. You can confirm the installation was successful by running the `git` command on the command line, which should return a list of commands and possible parameters to use.

Let's go ahead and navigate to one of the applications we previously deployed using the command line. Depending on where in your local computer you created the application, your path may look similar to this:

```
C:\Users\Username\Documents\GitHub\Machine-Learning-in-
Biotechnology-using-Python\chapters\chapter12\flask_cancer_ls
```

Go ahead and navigate to your directory either through a command-line window or by using **Visual Studio Code** (**VSC**). Once there, go ahead and **initialize** a new repository using the following command:

```
$ git init
```

Once the repository has been initialized, we will need to create a repository on your GitHub account (and later connect to it). We can do so using the following simple steps:

1. Log in to your new **GitHub** account, as illustrated in the following screenshot:

Sign in to GitHub

Username or email address

Password   Forgot password?

Sign in

New to GitHub? Create an account.

Figure 12.8 – GitHub login page

2. On the main page, click the **New** button on the left-hand side of the screen, as illustrated in the following screenshot:

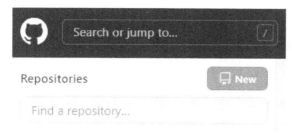

Figure 12.9 – Creating a new repository

3. Give the new repository a name, such as `flask-cancer-ls`. While leaving all other fields with their default values, go ahead and click **Create repository**, as illustrated in the following screenshot:

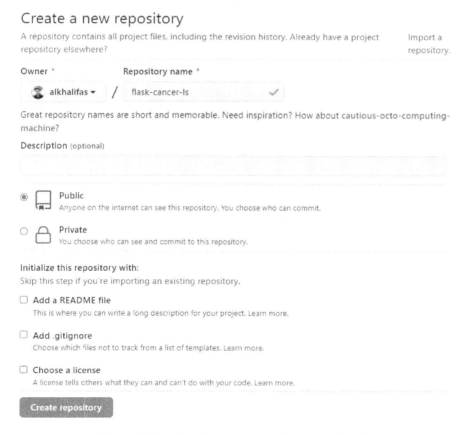

Figure 12.10 – Creating a new repository (continued)

Once created, you will be redirected to a new page containing some sample code for you to use. Given that we have already created a new repository, we will not need to go through this again. If we head back to the command line, we can go ahead and run the following command to add our files to `git` to be **tracked**. Tracked files are monitored by `git` and allow it to determine any new changes that need to be sent to **GitHub**. We could specify which file explicitly using the `add` command, like this:

```
$ git add app.py
```

Or, we could add all files using the period notation, like this:

```
$ git add .
```

It is generally considered better practice to add files individually as you are less likely to run into errors. Trust me!

With the files added to the *staging area* using the `add` command, our next step is to commit them. We can think of the staging area as a space where we stage our new code that is about to be sent to the remote repository. We can do so using the `commit` command, followed by a useful message describing the current commit, as illustrated in the following code snippet. You can use messages to briefly describe the changes in your commit. This will make things much easier when looking back at older code, trying to find a specific change on the GitHub website:

```
$ git commit -m "This is my first commit"
```

Upon executing this code, you may encounter an error asking you to specify your name and email address. Go ahead and complete this using the following commands:

```
$ git config --global user.email "you@example.com"
$ git config --global user.name "Your Name"
```

With your credentials saved and your commit completed, we can now go ahead and link together our **local repository** and the **remote repository** using the following command:

```
$ git remote add origin https://github.com/username/reponame.git
```

Be sure to replace `username` and `reponame` with your respective values! Once that is complete, you can go ahead and complete the final step, which is pushing your code to GitHub, using the following command:

```
$ git push origin master
```

With that step completed, if you navigate back to the GitHub website, you will be able to see your code here! Unlike the last two tutorials where we deployed our code in the form of an application to an online website available to end users to interact with, the objective here is to store our code and other content in a safe space. We have the option to allow other users to see our code or keep it private to ourselves. In addition, there are a number of platforms out there, such as Heroku, that are able to deploy an application simply by providing it with a link to the repository.

Summary

Over the course of this chapter, we reviewed a number of ways to deploy our application to end users in the cloud. First, we explored the use of AWS Lightsail, which allowed us to deploy our code in the form of an online web application using a Docker container. Next, we explored the use of GCP's App Engine to deploy our code, once again in the form of an online web application, using its user-friendly and abstract methodologies. Finally, we deployed our code in the form of a repository to GitHub, allowing us to expose the content to users, professionals, and tentative employers alike.

Congratulations! With this last tutorial now complete, we have come to the end of this book. Looking back at the last 12 chapters, we have covered many different topics in a diverse set of areas. In the beginning, we learned some new languages such as Python and **Structured Query Language (SQL)** and used them to analyze and visualize our data. We then explored some of the most common ML and **Deep Learning (DL)** architectures out there and used them to develop powerful predictive models. We then turned our attention to some specific areas of application, such as **Natural Language Processing (NLP)** and time series. Finally, we explored a few ways to deploy our applications to end users using AWS and GCP. Although we have covered a great deal of material within this book, there is still a vast galaxy of knowledge and information out there waiting for you. Before you move on to your next great adventure, there are three things you should always remember:

- The simplest solutions are often the best solutions. Never overcomplicate a model if you don't need to.

- Never ever stop learning. We live in a digital age where new discoveries are being achieved faster than ever.

- Metrics are your best friend. They will guide you throughout development and will help you make your arguments as a data scientist. Remember—everything is a sales pitch.

With these three things now in mind, go forth and do data science!

Index

Packt.com

Subscribe to our online digital library for full access to over 7,000 books and videos, as well as industry leading tools to help you plan your personal development and advance your career. For more information, please visit our website.

Why subscribe?

- Spend less time learning and more time coding with practical eBooks and Videos from over 4,000 industry professionals

- Improve your learning with Skill Plans built especially for you

- Get a free eBook or video every month

- Fully searchable for easy access to vital information

- Copy and paste, print, and bookmark content

Did you know that Packt offers eBook versions of every book published, with PDF and ePub files available? You can upgrade to the eBook version at packt.com and as a print book customer, you are entitled to a discount on the eBook copy. Get in touch with us at customercare@packtpub.com for more details.

At www.packt.com, you can also read a collection of free technical articles, sign up for a range of free newsletters, and receive exclusive discounts and offers on Packt books and eBooks.

Other Books You May Enjoy

If you enjoyed this book, you may be interested in these other books by Packt:

Bioinformatics with Python Cookbook - Second Edition

Tiago Antao

ISBN: 9781789344691

- Learn how to process large next-generation sequencing (NGS) datasets
- Work with genomic dataset using the FASTQ, BAM, and VCF formats
- Learn to perform sequence comparison and phylogenetic reconstruction
- Perform complex analysis with protemics data
- Use Python to interact with Galaxy servers
- Use High-performance computing techniques with Dask and Spark
- Visualize protein dataset interactions using Cytoscape
- Use PCA and Decision Trees, two machine learning techniques, with biological datasets

Learn Amazon SageMaker - Second Edition

Julien Simon

ISBN: 9781801817950

- Become well-versed with data annotation and preparation techniques
- Use AutoML features to build and train machine learning models with AutoPilot
- Create models using built-in algorithms and frameworks and your own code
- Train computer vision and natural language processing (NLP) models using real-world examples
- Cover training techniques for scaling, model optimization, model debugging, and cost optimization
- Automate deployment tasks in a variety of configurations using SDK and several automation tools

Packt is searching for authors like you

If you're interested in becoming an author for Packt, please visit authors.packtpub.com and apply today. We have worked with thousands of developers and tech professionals, just like you, to help them share their insight with the global tech community. You can make a general application, apply for a specific hot topic that we are recruiting an author for, or submit your own idea.

Share your thoughts

Now you've finished *Machine Learning in Biotechnology and Life Sciences*, we'd love to hear your thoughts! Scan the QR code below to go straight to the Amazon review page for this book and share your feedback or leave a review on the site that you purchased it from.

https://packt.link/r/1-801-81191-1

Your review is important to us and the tech community and will help us make sure we're delivering excellent quality content.

www.ingramcontent.com/pod-product-compliance
Lightning Source LLC
Chambersburg PA
CBHW081505050326
40690CB00015B/2933